T0220274

Nuking the World

Ecological Sustainability
and
Nuclear Power

**Critical information
about arresting "*climate change*"
and creating energy security**

International edition

By

Dr Bernardine Atkinson

© 2010 B.M.C. Atkinson
Director, Little Laurel Holdings Pty Ltd
GPO Box 1579, Darwin City, NT, 0801
Little Laurel Holdings Pty Ltd, ACN 095 032 447
Specialist Writing Services, NTBN 009 305 44
www.specialistwritingservices.com.au
1st Australian edition, June, 2009
ISBN: 978-0-9805767-0-2

Trafford International edition, July, 2010

Order this book online at www.trafford.com
or email orders@trafford.com

Most Trafford titles are also available at major online book retailers.

Printed in the United States of America.

ISBN: 978-1-4269-3355-4 (sc)
ISBN: 978-1-4269-3356-1 (hc)
ISBN: 978-1-4269-3357-8 (e-book)

*Our mission is to efficiently provide the world's finest, most comprehensive book publishing
service, enabling every author to experience success. To find out how to publish your book,
your way, and have it available worldwide, visit us online at www.trafford.com*

Trafford rev. 08/23/2010

 www.trafford.com

North America & international
toll-free: 1 888 232 4444 (USA & Canada)
phone: 250 383 6864 ♦ fax: 812 355 4082

NUKING THE WORLD

ECOLOGICAL SUSTAINABILITY
AND
NUCLEAR POWER

© 2010

Dedication

In loving memory of three outstanding women: my Great Aunt and mentor, Miss Phyllis Bryant, whose life's story quietly and steadfastly demonstrated the principle of *noblesse oblige;* my beloved, maternal grandmother, Sylve Davis (*nee Cain*); and my little sister, Kieranne Lucy Schwann (*nee* Atkinson).

Known to several generations of distinguished Australian women as *"Miss Bryant"*, Phyllis dedicated her professional life to the education of young women. She served as the Headmistress of Frensham Girls School, a Commissioner of Girl Guides and was a Life Governor of Winifred West Schools, Mittagong, New South Wales. Innovative, brilliant, generous and, in Godfrey Topp's words, *"wonderful to be around in an emergency"*, her integrity, honesty and vast organizational skills, helped to turn all with which she was involved into remarkable, high quality, undertakings. She was my personal mentor and best friend for several, formative, decades of my life.

Sylve Mary, was, simply, one of the world's very beautiful souls. Everyone who knew her loved her: she was accepting, gracious, generous, forgiving and thoroughly good fun!

Memories of Kieranne burn as brightly as the brightest candle on the darkest night for all who knew and loved her. Her heart condition meant that every day of her life she climbed an Everest or ran a marathon. Her courage, patience, fortitude and love for the people in her life, never waived, and her kindness and love-of-life helped to teach us to consider the well-being of others as being part and parcel of our own. That she endured continual *"prodding and poking"* under medical examination at regular intervals for the 43 years of her life helped the progress of medical science tremendously. Her congenital condition can now be rectified through an operation at infancy providing beneficial children with the expectation of a healthy life.

Acknowledgments

There are several people to acknowledge and thank particularly: Doctor Jim Mitroy, who opened my mind to the possibility of considering nuclear power as an option for Australia's future; Professor Helen Garnett, who provided particular encouragement at the onset of this communication exercise; Doctor Clarence Hardy whose advice and invitation to attend some of the deliberations of the Australian Nuclear Association resulted in this confirmed ecologist taking the nuclear option seriously and who patiently read and reviewed the entire first and second drafts of this book, providing some factual and editorial corrections. I would also like to thank my mother, Christine, and my father, Thomas, for their particular help and encouragement. For their help to enable my fledgling business, I am grateful for the professional assistance of Natasha, Rachel, Georgina, Nerida, Lisa, Cory and, in particular, Matthew and the team at Trafford. It is also *singularly* important to acknowledge Ricky Jermaine-Israel for his indispensable, personal support; whose genuine *"first impressions"* in response to the first section of this work enabled me to construct its contents more effectively. The content of this international edition has also been enriched by the occasional, research-rich, conversation with George Blahusiak. I am grateful to Noelene and Craig Isherwood and the authors of *"The fight for an Australian Republic"* for their permission to quote from that important story. Finally, I would like to especially acknowledge Professor Leslie Kemeny's encouragement and gracious allocation of precious time to review and contribute to my work. His distinguished career has presented him with many accolades, but in the Australian press, he is known as *"Australia's best scientist"*. Professor Kemeny's particular encouragement has amounted to an immeasurable endorsement. It has been my absolute privilege to have worked in this small way with such a remarkable fellow.

To the above-mentioned individuals, please accept my most formal, and sincere, *"thank you!"*

Synopsis

While the world wallows in fear about the rapidly manifesting effects of global climate change and muddles about advocating a plethora of alternative energy sources for the future, here the immediate transition to nuclear power is unhesitatingly advocated. To do so with authority and conviction the myths used effectively to "*spin*" confusion, and create fear, about the very word "*nuclear*" are exposed and unraveled. Then, when nuclear power development is contextualized against the determinants of genuine ecological sustainability, one single, confident, result emerges. *Ecological Sustainability and the Nuclear Power Story*, first written as a series of letters to the Australian people and now appearing here as an international edition, hoping to very positively and safely "*Nuke the World*", unequivocally **ends** debate on the import issue of determining sustainable energy alternatives for the world. Nuclear powered electricity generation is "*the way to go!*"

Dr Atkinson's first letter investigates the eight significant lies used by the unscrupulous and ignorant to withhold and delay the empowering technological revolution that nuclear science bequeaths to all humankind. The letter's detailed investigation concludes that nuclear power, now a mature, safely-managed, technology, is able to provide inexpensive, electrical energy and desalinated water, to the entire world, for thousands of years.

The second letter looks seriously at the phenomena of "*Climate change*". It emphasizes the important role forests have played in ameliorating climatic extremes to create stable climatic temperature thresholds. Portrayed with inaccurate and interminable complexity by the media, to halt and reverse climate change is actually quite simple: fossil fuel burning and other polluting activities must cease; extra-somatic energy consumption must be reduced; and, critically, amongst other life-sustaining practices, the world's forests must be replanted. Also provided in this letter are clear, logical and original arguments that describe why nuclear power is the best base-load alternative for reliable energy production when assessed according to five significant parameters. These criteria are called the *"quintuple bottom line"*. When, for example, dimensions such as *"energy conservation and efficiency"* and *"resource-use efficiency"* are added to the policy-makers' *"triple bottom line"* of qualitative developmental determinants, nuclear power emerges triumphant on every front. Not only that, mini, mobile power generators and *"modular"* desalinators will provide power and pure water to enable remote communities to strategically reforest and practice secure agricultural and horticultural production in areas now abandoned as wastelands. Permanent stream-flows can be re-established and the hurting biosphere, buffeted by more than two centuries of unfettered, industrialized exploitation and several thousand years of sustained onslaught against its forests, can be repaired by humanity asserting its collective determination to exercise *"wiser dominion"* and stop *"befouling our nests"*.

The third letter remains exclusive to the Australian edition. Here, a fourth letter in the series is presented to discuss the implications of Australia's political abandonment of the nuclear power alternative from 1970 onwards. Australia possessed more than 30 percent of the world's most accessible, concentrated, uranium ore and another ten percent in less accessible and poorer quality ore assemblies. This ore is rare, finite and invaluable. Nuclear science-sophisticated nations, with no uranium ore, have benefitted from the obfuscation of the ore's true value and been allowed to extract this irreplaceable resource from Australian ownership and management for nearly a century. That exploitative time of politically-endorsed, offshore, opportunism is now over and Australians can commence to reap the benefits of the resource for themselves and their Oceanic neighbours. Nuclear power, instead of being regarded fearfully as a source of trauma, war and disunity, can now be more wisely applied to create *an incomparable millennium of global prosperity, and replenishment, providing us with the commensurate expectation of peace and security.* This vision's realization requires us all to implement the *"meritocratic"* ideal of wise, benevolent, political, leadership and embrace the *"quantum leap forward"* opportunities nuclear-powered development provides for all humanity and our precious planetary home, Earth.

Dr Bernardine M.C. Atkinson

29ᵗʰ April 2010

About the author

Dr Bernardine Atkinson is an ecologist and a communication specialist. Dr Atkinson read for a Bachelor of Science at the Australian National University then completed an Outdoor Education qualification at Kelvin Grove College of Advanced Education, after which she began to specialize in communicating ecological concepts. Her Master of Philosophy research degree was conducted at the Centre for International Journalism Studies, College of Cardiff, University of Wales. Her research then identified *"climate change"* as a more appropriate analogy than the *"greenhouse effect"* and she devised a *"new"* news formula for journalists.

On returning to Australia she developed the world's first *Ecological Reporting* course for Deakin University's, then, School of Literature and Journalism. Other important work included devising several successful Landcare strategies to enrich biodiversity and protect freshwater in Western Victoria. More recently, her doctoral studies in strategic communication investigated and devised strategies to accelerate the transition to ecological sustainability.

Dr Bernardine Atkinson was honoured for her work in Britain by being nominated a 21ˢᵗ Century Environmental Trust Fellow.

Contents

Appendix one

Preamble

Unraveling *"spin"*

"Spin" is raveled from ignoring or distorting truth.

An unfailing, important rule for effective debating is the need to have agreed and precise definitions applied to the subject under discussion. If that which is *"renewable"* is actually only *"intermittently renewable"*, i.e. discontinuous, inefficient, incredibly expensive and dissolute (*wind, solar and geothermal*), but presented as *"sustainable"* by one side of the debate, then the whole effort to discuss the subject of *"renewable and sustainable energy"* becomes nonsense. Propositions held to be equal are not equal. In the global debate about future energy alternatives, not only have definitions been butchered, the fundamental premises supporting the solar, wind and geothermal lobbies' *"renewable is sustainable"* arguments, are often incorrect. The sustainable energy security required to support modern urban communities and industrial applications can only be **partially** met by inefficient, intermittently-renewable sources. The requisite energy security **is** attainable only when we have high conversion efficiencies and continuous, high-grade, renewable sources of supply. Such reliability is able to be provided by hydro-electricity (*which most of the world does not have in abundance*) or nuclear power. And here we face a substantial challenge: providing nuclear power to the world requires the global community to achieve trade relationships that provide an equitable distribution of the enabling, but incredibly finite, uranium ore resource **and share** the associated technical knowhow and managerial skills for its safe use.

The second inviolable rule for establishing the truth of competing propositions is to have *logically consistent* argument. If the intention is to reduce carbon pollution but the *"schemes"* devised to facilitate our transition to new patterns of energy production *actually subsidize* polluters to continue polluting, then the scheme fails this second test. The proposition's logic is not *"internally"* consistent.

As a consequence of these and similar distortions, sensible, public debate around the globe about the nuclear power alternative is disabled: garbled, contradictory nonsense and misinformation prevail, i.e. *"spin"*.

The presentation of fact-based research, such as is sought to be conveyed in this book, is the only mechanism by which we can hope to elevate *"popular, ill-informed, opinion"* to the more enlightened position of *"informed opinion"*.

To date, the only significant criticisms of the thoughts proffered here have come from politicians who have said that it is too *"dissenting"* and a potential publisher who said that the information challenges the prevailing beliefs to the point of creating *"discomfort in the minds of its readers"*. But, without the expression of *"dissent"*, we have no way of determining the merit of competing propositions and, thus, no hope of progressing from ignorance to enlightenment. The progressive benefits of truthful, unfettered, communication are the fundamental reason for the enshrinement of *"freedom of speech"* as a universal value and foundational *"human right"* in all progressive cultures.

The policies developed in Australia over the past forty years have ignored the opportunities that nuclear power would provide to our nation and our world: these decades have not evidenced a *"meritocratic era of governance"*. Fortunately, democracy itself is an evolving institution and, in the near future, I am sure we shall find some way of ensuring that non-covetous, super-intelligent and wise people, rise to leadership: people prepared to assume responsibility for creating and implementing the forms of governance which will help the global ideal of democracy elevate towards the *"meritocratic state"*. A *"meritocracy"* is an advanced form of democracy that specifically seeks to assert wise political judgment, not simply follow the simplistic and often incorrect thoughts and beliefs of the majority of voters. (*This idea was shared with me by an eminent Australian musician, Ian Ellis. Mr Ellis constantly sought perfection in musical performance.*) Such natural, political, evolution cannot happen if we quash the expression of "*dissent*".

Furthermore, intelligent leaders do not fear criticism: they welcome it as an opportunity to test and improve their perspectives and policies. When constructive criticism of government policy and alternatives, such as nuclear power, are able to be fairly portrayed by an independent nations' media (*without those who convey the message grappling with the risk that their employment could be terminated*), and results in more sensible policies emerging, we shall then know that some of our social edifices have achieved the ability to organize for governance by fearless and intelligent leaders. We now only oscillate about this potential being realized in Australia.

Prosperity plus

Before all of humanity lies the most amazing era of prosperity imaginable: a millennium of replenishment, plentiful energy, abundant pure water, and full employment: all possible when we carefully, and strategically, implement the amazing opportunities nuclear power provides for both electricity generation and desalinated water production. To enable this transition, the unconscionable rhetoric used both to manipulate and manufacture public perceptions on the issue of nuclear power adoption must be exposed for that which it is: incognizant diatribe. Real knowledge must replace "*doublespeak*" and innuendo. False reasoning must also be exposed. Fortunately, the genuine determinants of ecological sustainability are now known. And, as this book commences to show, nuclear science is not an incomprehensible science for most people, so we have the hope that our transition to genuine ecological sustainability will not take long.

We must move speedily because hungry human activities have nearly resulted in terminating the habitats that support all advanced forms of life on this planet. Massive extinction will certainly happen if we do not halt and reverse the degradation of the biosphere. Humanity is capable of assuming responsibility for remedying this destruction. Unshackling ourselves from the selfish, exploitative, tendencies that have caused massive disruption and destruction to the living biosphere is the challenge we individually and collectively must win. The battle lines are drawn: caring co-operative edifices versus self-serving, inconsiderate exploitation. Only the former is empathetic to the natural co-dependencies that sustain life: all life.

We will achieve this transition challenge. Why? Because, not only do we have no choice - our collective survival is at stake - but also because, within our inherent human nature, the propensity to solve problems is much stronger than our desire to create them. Problem-solving is one collective trait of which humanity can be proud. Most of the ecological problems we now face resulted from our blithely implementing industrial mass production and land-clearing practices without understanding their consequences. We now know, and have identified, unambiguously and correctly, the causes of our ecological malaise, so we really have no excuse not to, energetically, and joyfully, commence to rectify these matters.

Quantum physics and quantum mechanics are part of the fruit of our problem-solving propensity. Nuclear science, and the quantum physics upon which it is based, provide the most sophisticated, accurately predictive, and objective, knowledge of the material world yet achieved by humanity. Some pioneering individuals in the scientific community (*Bohr, Planck, Einstein, Rutherford*) reached these insightful, intellectual pinnacles of thought more than a century ago; others have know of these possibilities for at least the last sixty years (*Fermi, Oliphant*): the rest of us are now about to catch up!

The prevailing global belief in the current, internationally-pervasive, political prognosis of economic and ecological gloom, will change overnight when the world commences to, intelligently, implement nuclear power, globally. Quantum science's revelations and our ecological insights bequeath the dawn of a wonderful new era for humanity: the nuclear epoch and a millennium of replenishment.

A nuclear-powered epoch will enable freshwater scarcity to become a relic of the past. Careful application of this technology means that deserted, abused, landscapes can become productive and green. The salt-heavy waters supplying artesian-well dependent places, now limiting the opportunities for life-on-land, can be desalinated and, where under-ground resources are exhausted, they can eventually be replenished using water extracted from our oceans; coastal cities and regional centres can trickle fill their reservoirs without waiting for hoped rainfall; trenches carrying ocean water could continually pass through condensation chambers allowing inland lakes to be replenished; exploited river flows could be restored along their length and breadth through man-made, forest-filtered, trickles of excess fresh water produced from strategically-placed, mobile nuclear-powered generators and desalinators, enabling thriving biological and economic activity.

By strategically reforesting and planting valuable shrubs and trees for their fruits, leaves, seed, flowers, shelter and wildlife habitat-creation potential, humanity will be, not only, able to provide genuine wealth through the provision of useful fibrous materials and food and building materials, we will be commencing to halt and reverse the climate change induced by excessive carbon dioxide. Planting trees and shrubs *en masse* will eventually clean our globally-poisoned atmosphere thick with sulfur dioxide, sulfuric acid, nitric oxide, coal tars, and carbon particles, and release oxygen and water vapor, helping to restore health to the air we breathe.

Nuclear-powered electricity *"is the way to go!"* Presto! Suddenly, our precious planet's future is looking positively and amazingly prosperous.

Invaluable ore

As you read through the information shared in this book, you will discover that concentrated uranium ore is invaluable: it has the potential to provide the entire world with energy security for many thousands of years.

Most of the world's irreplaceable uranium ore bodies have been accurately identified by aerially analyzing comparative, emitted radiation levels. The richest, and most accessible, supplies have already been mined and placed in storage. France now has 100,000 tonnes of *"yellow cake"* (*uranium oxide*) in storage. In this, they have safe, clean, fuel for electricity generation for themselves (*and the countries in Europe from which they currently earn billions of dollars from each year for supplying their electricity*) for at least five thousand years (source ANA Newsletters, 2008/9). How can anyone place a value on such an enabling resource? It is priceless. The only commensurate, recompense-able value able to be returned to the mined nations for the provision of the ore would be a royalty apportionment of the mined resource (*perhaps, 30 percent*) and access to the enabling knowledge to allow them to usefully and safely apply this remarkable technology. An apportionment retains its value, regardless of currency value differences and market vagaries: and it provides for economic capacity building that is foregone if a mere cash-royalty or tax is accepted for its exchange. National leaders would be wise to institute this form of apportionment not only for uranium, but for all of Planet Earth's natural resources. If we do not, it is highly likely that the resentment and impoverishment continued resource exploitation will incite, will remain to make the entire world an unhappy, defensive, terrorist-traumatized: an unhealthy place in which to live.

Why is this ore so valuable? Uranium doesn't burn and produce flames like logs in a fire would. Rather, when it degrades (*transmutes*) into other substances, it emits light and energy: a great deal of energy. Gram for gram, uranium produces many thousands of times more energy than would be achieved from burning either coal or oil.

The facts are staggering. Uranium's energy output, per kilogram of fuel, is 500,000 mega joules; crude oil burning gives us only 45 mega joules per kilogram; natural gas provides 39; while brown coal provides 9 mega joules per kilogram (*Australia's Uranium*, 2006, pp151, 152). When the *natural* isotopic changes that commence uranium's "*burning*" are captured and controlled in a reactor vessel, by using just a tiny, little, bit of "*yellow cake*", we can achieve the same result from **one ounce of mixed oxide fuel** (*uranium and the even denser, reactor-made element, plutonium*) as we would achieve by burning **more than one million ounces of coal** (*Professor Kemeny, pers. comm. 2009*).

There are other benefits: burning uranium in a reactor results in **zero** carbon dioxide emissions per kilowatt hour. Burning gas, coal or fossil fuel oil, all result in 800 plus grams of carbon dioxide per kilowatt hour. In the remainder of its fuel cycle (*transporting and processing the ore*), nuclear power contributes only 5 grams of carbon dioxide per kilowatt hour; but fuel-oil contributes 149 grams; coal, 111 grams; and natural gas, 68 grams. While manufacturing photovoltaic, solar, panels, contributes 97 grams of carbon dioxide per kilowatt hour (*Australia's Uranium*, 2006, p168).

Nuclear power provides even more good news. We can convert uranium's energy, super-efficiently - that is with little or no waste - to electric power. Nuclear-powered electricity production is tremendously efficient: it is physically possible to achieve 90 percent plus conversions. Solar and wind power can convert *maximums* of 25 percent of their received energy into power ... *sometimes*: when the wind blows and when the sun shines and reaches its zenith; most actual conversions are much less. These technologies are not energy-conversion, efficient, processes.

There is more satisfying news. A nuclear reactor's energy-capturing technology is now so well known, and has been so carefully developed, that it is, without contest, the safest way to generate electricity yet devised. The processes now used, in heavy and light water pressure reactors, leave no carbon footprint and, eventually, in stark contrast to that which we are told hyperbolically by pseudo-scientists, it is possible to recycle *all of the fuelrods'* High Level Waste and use it to produce more electricity. In truth: the word, *"waste"* is a misnomer and another excellent example of the prevalence of the *"spin"* pertaining to all matters nuclear.

The global, High Level Waste disposal quandary has always been *"a temporary, storage issue"* (*Colin Keay's words*). Even using the technology we now possess, 97 percent of a fuelrod - its elemental metals - can be dissolved and separated, then recycled for reuse as fuel. This sort of material, incorrectly called *"waste"*, while it remains hot, is now safely stored, on-site, at many nuclear facilities all over the world. All of this material will eventually be recycled many times over. Reprocessing *"spent fuel"* happens today in Britain (*Sellafield*), Russia (*Chelyabinsk*), France (*La Hague*) and in Japan, the United States, Germany and Switzerland.

The residual three percent of a spent fuelrod, the "*ash*" produced from the permanent transmutation of some of the uranium into lighter, unstable, elements, can, when responsibly and carefully managed, be safely disposed of to present no hazard whatsoever to humans or the living biosphere. The process emulates nature's way of preventing surface contamination: it is called "*deep geological burial*".

You might not yet agree with all that has been expressed here, but after reading the information that follows, I am sure that you, too, will be convinced that Planet Earth's citizens would be wise to **immediately** commence to plan for the safe implementation of nuclear power for all the nations of the world and, in doing so, enable our transition to genuinely-sustainable ways of living.

Letter One
Nuclear power matters!

Dear Friends,

I have not known how to alert you to an amazing and embarrassing manipulation of our world, the nation of Australia in particular, relating to the issue of nuclear power generation, except by putting pen to paper and using the *"freedom of the press"* to express this voice of genuinely *"disinterested support"* for the careful, global adoption of the nuclear power alternative to fossil-fuel burning power generation. This information was first published in a series of letters for the Australian people to consider, but the information is so important that we have felt compelled to assemble a little book for international distribution as well. The information provided here is intended to help to commence more informed discussion about nuclear power and to accelerate its adoption world-wide. It is also designed to help enable all manner of local decision-makers to improve their planning on all issues pertaining to local, national and international energy security and ecological sustainability and as such, it is particularly mindful and respectful of the critical imperative all human beings now have to restore ecosystems, particularly bio-diverse forest biomass. This urgent restoration is required to help our planet resume its natural propensity for supporting the further evolution of life. Bio-diverse biomass is able to modify climatic extremes: a function provided by the living world's remarkable propensity to provide and achieve health-maintaining, physical systems equilibrium through its myriad of interdependent, mutually-sustaining, relationships.

Myth dispelling

To commence: many millions of people have been hugely misled on this issue: **nuclear power is safe, clean and sustainable, eminently manageable and the small amount of waste *can be* disposed of safely.** If we replace the world's archaic electricity-generating utilities with modern, nuclear power production *and concurrently apply* ecologically sustainable development principles (*particularly bio-diverse, biomass replenishment and energy consumption conservation*), electricity will cost much less than the current cost; coal will not be burnt but will, instead, be used to create both synthetics and the polymers required for hydrogen gas containment; consequently, tremendous, new, sustainable industries can be developed. Our planet's future is, assuredly, incredibly bright and prosperous if we, united and globally, come to grips with not only the substance of our having been duped on nuclear matters, but also immediately commence to harness the opportunities it provides for enabling economic prosperity.

We, except for the few who have seriously studied this issue (*and too often they have been either public and civil servants unable to speak openly because of the confidentiality clauses associated with their employment, or nuclear scientists who usually only communicate with their peers*), have been misled on several significant fronts. I am inclined to call them *"The Eight Lies"*, but in fact, these eight, primary, sources of misinformation, are misrepresentations or they represent half-truths, distortions, misconceptions or myths.

Firstly, we are told that radiation of any form is harmful to our bodies and will cause cancers and mutations.

Secondly, we have believed that the uranium resource is finite (*a fifty-year mine-able commodity worldwide*) and so nuclear power is not a sustainable energy option.

Thirdly, depleted uranium, uranium-238, is a waste product, only 0.7 percent of the uranium oxide is valuable: the unstable uranium-235 isotope.

Fourthly, we are advised that nuclear power is **so** dangerous it cannot be managed.

Fifthly, we are told that uranium enrichment produces plutonium and plutonium is feedstock for nuclear weapons and so we should be terrified (*and we are*), and this fear has petrified debate on the adoption of nuclear power.

Sixthly, we are told that a nuclear power plant could explode like a bomb and that the Chernobyl hazard remains in every power plant.

Seventhly, we are told that waste material is unmanageable.

Eighthly, we are told that plutonium and uranium will emit dangerous radiation for hundreds of thousands of years and so will be forever hazardous.

Let me assure you that all of these positions, one to eight, are either false or represent a significant distortion or omission of truth. Entire nations have fallen prey to what would appear to be a combination of general ignorance, misinformation and deliberate propaganda. The reason for the duplicity is that uranium is, not only, of inestimable worth, it is invaluable. There have been smart, vested, interests making the truth about nuclear power matters totally obscure. Many nations, especially those able to supply uranium ore, have, very probably, been subjected to deliberate propaganda, designed to create and perpetuate fear of anything nuclear, to ensure that this incredibly rare resource is able to be acquired, inexpensively, by the few, well-informed, nuclear nations that possess little or no uranium ore.

That exploitative time is now over. Our world is now freer to become more caring and more enabling of the growth of all nations, especially those that have possessed raw resources but have had little technological knowledge. By sharing this information, we have the expectation of humanity becoming more responsible, more equitable and, as a consequence, far more responsible, healthy and happy in its entirety.

The nucleus of the nuclear power story

The technologically-advanced nations of the world know nuclear electric power to be "*safe*". In fact, it is the safest (*and cleanest and, rapidly becoming, the cheapest*) method of generating electricity yet devised. But, a huge proportion of people have feared this development. At the root of the widespread, wrong, opposition to nuclear power, we find one fundamental commonality. Many people have been frightened of "*radiation*" and have associated nuclear electricity production with nuclear bomb production and the horrific radiation burning and destruction that can result from the detonation of such appalling weapons of mass destruction. As a consequence, nuclear power has been banned in places like New Zealand and Australia where, instead of people being well-informed, a rather pompous, self-righteous, ignorance about nuclear energy issues has prevailed. Many of the "*eight lies*", mentioned above, have been promulgated and remain entrenched in such nations'. They have been used to defend what are, essentially, unsustainable policies pertaining to energy infrastructure planning and development.

Before commencing to examine, in detail, each of the eight foundations of nuclear myth-making, it is important that we absorb some fundamental nuclear science ideas. We can easily understand nuclear science if it is presented to us in English and not in complex, conceptually-rich, chemical and mathematical formulae. Those, scientifically-accurate forms of expression are necessary for its engineering and technical applications, but not for general understanding.

While I am not a nuclear scientist, I am a scientific thinker and an ecologist, and I thought it helpful if I share my own rudimentary understanding of the atomic insights gleaned from the progress humanity has made from its last one hundred years of passionate, purposeful, scientific enquiry. I hope you will not be daunted by this preliminary exercise in scientific interpretation. It is the toughest section of the book for a non-scientist to read, but if you manage to absorb the following, fairly rudimentary explanations, the preliminary understanding gleaned might help to launch your own more substantial, personal enquiries into this most interesting and conceptually-rich, discourse. You will discover that what *"goes on"* inside a nuclear reactor is really not a scary process after all, especially when the facilities are soundly engineered and the operators implement careful managerial precautions to avoid any unnecessary, and all excessive, radiation exposure.

Let us commence with an explanation to defray the irrational phobia so many of us seem to have developed for the words *"radioactivity"* and *"radiant energy"*.

Radiant energy

Humans emit radiant energy (*some people can see this emanation, called Kirlian photography, and call it our "auras"*). We are also surrounded at all times by various radiations of different intensities. Radiation impacts our bodies continuously from the air and from the Earth. All volcanic activity and the hot geysers of water that spout forth are produced by radiant energy emanating from the Earth's molten, dense, and slightly unstable, metallic core. Our closest star, the Sun, is one of countless billions of enormous nuclear radiators. If its energy was not available, life, simply, would not be possible, but we also know the Sun's energy, and that produced by molten lava, must be *"moderated"* to be of use to humans. Too much intense radiant energy can hurt us. If we jump into hot lava, we would quickly become less than cinders. So, humans strive not to live close to volcanic activity. If we stand naked, all day long, in the Sun's hot, tropical, radiant energy, it is likely we will suffer sunstroke. So, to live safely within the ambit of the Sun's intense radiant energy we either, cover ourselves with a thin layer of clothing and wear a hat, or, we minimize the time spent in the sun and find something to provide us with shelter.

These are exactly the same sort of strategies we use to make the comparatively small amounts of radiant energy produced in nuclear reactors "*safe*". We manage and moderate the radiation from the metals in a nuclear reactor to **minimize** our exposure to radiation to a level similar to that which we receive naturally. We also design structures (*from lead, concrete and steel)* to safely contain and confine human-made nuclear reactions and we continue to refine our engineered structures and fuelrod manufacturing to ensure that human-, and biosphere-safety come first. With precautions of this kind, men and women have lived near, and worked daily in, nuclear power facilities, without harm, for the last half-century.

Besides our inherited, ancient, understanding that all Creation comes from God, we now have the rudiments of scientific support for understanding our Universe's origins. In the ancient Creation story, we know that God, the Creator, the source of all light, order and, hence, all energy and its processes, gently breathed a word into being and, in so doing, scattered "*cosmic dust*" into the formless, but form-receptive "*void*" (that "*ether*", "*the elastic, subtle substance believed to permeate all space*" *[Shorter Oxford English Dictionary]*, which, though it appeared formless, remained full of receptive-potential for material manifestations) and so set the creation of our material universe into being. Some of scientific understandings compliment this ancient Greek wisdom and although scientific, fact-based, explanations have emerged only slowly, they now provide predictive and precise insights never before attained by humanity. One branch of such scientific thought has been tremendously helpful in this respect: Nuclear, or Quantum, science.

We call the study of the centre of the atom, the study of its *"nucleus"*, hence, *"nuclear science"*. *"Nuclear science"* is also often called *"quantum physics"* a term derived from our study of the discreet parcels of energy, *quantities* of energy that, when coalesced, form the particulate *mass* of the coherent, material entities we call *"atoms"*.

Perceptive Einstein deduced that material matter and energy (*which manifests as "light and heat"*) are extricably interchangeable: energy equals mass *times* the speed of light's speed meaning that stationary mass can de-compartmentalize to become both moving energy (*heat*) and light (*radiation of various intensities*) and move in a wave-like fashion. Light energy can also operate as distinct *particles* of energy. Particles of energy can manifest as mass and mass can manifest as energy; and mass, when it *changes* into other forms of mass, both absorbs and emits waves of energy.

When light photons (*particles of energy measured as "quanta", i.e. particular quantities of energy*) are *"constrained"*, they cause matter to materialize and to alter form. *Mass* can therefore be described as *"constrained and contained, energy"*.

So, rather than being the antithesis of matter and life, radiant energy is inextricably part of all matter, and so, it, in turn, forms the basis of the material constituents of life. In fact, *"radiant energy"* of different intensities is the *"operative force"* behind all the processes of our material universe because all elements are made from varying quantities, proportions, structures and configurations of the same basic ingredients: protons, neutrons, electrons and constrained, energy.

"Life" itself appears to be a precious, self-determining, precisely regulated, sophisticated, organization of energy and coalesced energy: mass. But even so, our understanding of life and the

emergence of consciousness remains severely limited. Perhaps that which we consider to be "*inanimate*", being made of the same fundamental components as that which is "*animate*", may possess a measure of consciousness and self-determination that we are sensorially oblivious to? If this were so, surely we could begin to make sense of the ancient Scriptural wisdom that says, in response to the majesty of God, "*even the rocks and stones themselves shall start to sing or find voice!*" (Chronicles I, 16:33, Psalm 65:13, Luke 19:14). The pattern evident in the *Story of Creation* is that the material world evolves into ever-more complex, mutually-sustaining, organizations and component complexity, hence, more creation. When self-determination and self-perpetuation reaches the point of complexity to allow memory and self-awareness to manifest, we have "*consciousness*".

"*Consciousness*" itself may illustrate progressive levels of complexity from the obtuse to self-reflective, then sublimely sensitive, higher orders of being, then super-refined, transcendent or spiritual being. Humanity, with the ability to describe its own states of self-awareness, represents a pinnacle point in this creative journey.

Movement which is *deliberately* discordant and without internal harmony (*the gross manifestation of this being represented by the idea of* "*chaos*") results in destruction and the de-fabrication of ordered complexity: a reversal of Creation's impetus towards greater integrated, complimentary-complexity, manifesting.

When isolated and reduced, the sub-atomic world's particulate, distinctive, elemental, arrangements can emit particular quantities, *quanta*, of radiant energy in the form of alpha, beta and gamma radiation.

Very simple, sometimes miniscule, combinations of attraction and repulsion forces cause these sub-atomic energy parcels to bind together *intensely.* Opposites attract, like-forces repel. Ironically, it would appear that the tinier the operative, attraction *(or repulsion)* force is, the stronger the bonding *(or the stronger the repulsion)*. The resultant forces that bind and bond an atom's biggest sub-atomic particles, protons and neutrons and their outer clouds of whirling electrons, are very strong: they are the glue that hold the material universe together.

When protons and neutrons and electrons bind together, mass manifests, at first, in the form of an elemental atom, then many atoms of the same element, then combinations of elements. When the bonds between a single atom, or combination of atoms, are created, or when they are loosened, or break, *radiant* energy *(rays of energy)* is released. We capture this energy in nuclear reactors to turn water into steam and turn turbines to generate electricity.

Different-sized particles of different energy intensities, moving at different speeds, in close or far proximity, create a variety of forces, including gravity, kinetic energy, and magnetism's attraction and repulsion. Just as a slow-moving pocket of water is turned into a savage whirlpool by an unimpeded, fast-flowing stream running beside it, so, proximities alone create movement and, in like fashion, the interactions between different energy intensities and movement, impeded and unimpeded, provide for all the known forces of the universe to manifest. Even the tiniest manifestations of these forces can cause huge deflections and changes in movement and momentum.

The jagged fragment of an exploded remnant of star tumbles into spherical shape with the heaviest elements concentrating in its centre; from the spin that develops around the sphere's poles and its relation to other orbiting entities, gravity and tidal forces come into being and the physical characteristics of a life-supporting planet are eventually coalesced: the lightest elements forming atmosphere; the denser remaining at the planet's internal core.

Thoughts like these are how I personally reconcile ancient understandings (*that the entire universe was created from a single word of God),* with the insights of modern physicists. In this way, we can imagine how just a tiny stream of particle disturbance from a minute quantity of cosmic dust from tiniest breath of God, purposefully spoken into the, full-of-potential, void, could eventually create an entire universe!

Radiation-emitting atoms and elements

When different *elements (an element is material which cannot be resolved into a simpler, stable, material manifestation)* are formed (*by energy organizing itself as different quantities of mass*), they each represent a *most* particular *pattern* and arrangement (*configuration*) of tiny, tiny, sub-atomic forces and particles, each unique element with a slightly different quanta of energy and, subsequent, unique configuration.

In many instances, these element-specific, unique, sub-atomic-particle arrangements are perfectly balanced at the pressures and temperatures that prevailed at the time of their formation. But some of these element's atoms, when forming, were always left with one or two (*mathematically predictable*) imperfectly formed, slightly unbalanced and unstable, embedded, similar atomic arrangements. These are the left-over material that did not contain quite the right "*quantum*" to make a stable form of the atom emerging. These similar, but "*short-changed*", atomic arrangements of protons, neutrons and electrons we could describe as the "*unstable isotopes*" of an element. They are also called the "*radioactive isotopes*".

This process might be easier to understand if you imagine an atom to be like a ball and tried to squeeze a "*group of nine balls*" into a container and found they fitted perfectly, and so created a unique and stable element. But when you put a *"group of ten balls"* into the same container: yes, they could all fit, but the ten balls might each assume either the same, slightly different shape to the nine as a consequence of being *"squashed"* into a particular space (*and if their new shapes were identical and stable, we would say a new, stable, element was created*). Or, only one of the ten balls might be *significantly different* shaped having been *"squashed"* while the other nine atoms retained their original, integral, shape. In this last arrangement, the nine *"ball-like"* atoms that retained their same, internal arrangement, we would call *"stable"* atoms, but the *one* which changed its composition slightly, but still remained one of the group, we could regard as an *"isotopic"* manifestation of that element, a short-changed or *"unstable"* atom of the element.

On our planet, more than 70 of more than 100 known natural elements have naturally-occurring radioactive isotopes. All isotopes have varying levels of energy-mass stability: some are very nearly as stable as an element; others might have a life of only a few seconds before emitting the internal constituents that make them *"top-heavy"* and teetering in an inherent quest for stability. The internal material they emit to achieve internal stability is expelled as excess energy: *"radiant energy"* and sometimes radiant, sub-atomic particles.

Eventually, in nature, all *unstable* atoms will, over time, rearrange their internal composition to become *"stable"*. They may emit a neutron, or neutrons plural, or expel protons, neutrons and electrons, and so, intense parcels of energy (*alpha, beta and gamma radiation*) are released: we call this process atomic *"decay"*. The energy emitted from this natural instability, happens rapidly or very, very, slowly, and we call it *"radioactivity"*. Radioactivity is completely natural.

Hydrogen, carbon, cobalt, nitrogen, gold, silver, and copper all have a few, unstable, radioactive atomic shapes or *"isotopes"* happening in their internal atomic assemblages. So does the element we are particularly interested for generating nuclear power: the heavy metal, uranium.

Uranium is the heaviest natural element because it is very densely packed with hundreds of protons and neutrons. Like all unstable isotopes, the unstable, uranium atom has the capacity to release a *great deal of energy from the neutrons it expels in its inherent quest for stability, at a predictable, known rate.* In pure uranium, only a tiny, little, part of the elemental patterns that form uranium manifest as the important, unstable, radioactive isotope, uranium-235: less than one percent. 0.7 percent to be more precise.

Over time, and in the inherent quest to rebalance and stabilize its internal constituents, the unstable, radioactive, uranium isotope eventually emits enough internal particles to become the slightly lighter, stable element, lead. Fortunately for us, because it does so at a very predictable rate, we have been able to use the energy released in this natural decay process to generate heat which we can capture **entirely** in a reactor vessel and then apply to create electricity.

But by confusing a nuclear reactor power generator with an atomic bomb, we have succumbed to a grossly distorted idea of what actually happens in the reactor-controlled, nuclear fission/fusion processes. The condemnable atom bomb is designed to cause billions of atoms to release energy **simultaneously**. Nuclear power generators are entirely different. In man-made nuclear reactors, isotopic uranium's naturally emitted particles and energy are encouraged, through their cleverly designed containment, to combine with a small number of *stable* uranium atoms, and so, in turn, make them unstable and the ensuing, contrived, instability in new atoms, causes the neutron-receptive material to emit more internal sub-atomic particles and, in so doing, release more heat energy.

This *"nuclear re-action"* is so named because the *"nucleus"* (*the centre*) of an unstable atom *emits* radiation (*generally excess neutrons*) which can cause the atoms closest to it to absorb energy and, as a consequence, change their own internal nucleus' composition and become *"unstable"* in turn: the stable and unstable material have *"re-acted"* together (*hence the name, "nuclear reactor"*).

This process, when activated by interspersing small amounts of unstable uranium alongside stable uranium, creates a chain-reaction of instability and, as a result, slowly commences to produce heat (*after about three days*), then a great deal of heat. It is able to maintain heat production until the process is naturally slowed or stopped, eventually, by the production of non-reactive, stable isotopes and elements.

These deliberately-induced, staggered, slow, nuclear reactions are contained in fuelrods placed in an impervious, steel *"reactor"* vessel. I have held an Australian-designed fuelrod in my hands. Of course it was not a *"used"* fuelrod: that would be too hot to handle and would require water immersion for a number of years to allow the *"burnt"* (*atomically-altered*) uranium metal to cool down. But, like volcanic lava, it does cool down and, if it is surrounded by water to absorb excess neutrons, the fuelrod can be recycled many times, until all of the metal's potential for instability is used-up.

The fuelrod I held was a 60cm long, slender, rectangular steel canister, with a width and height of about 10 cm. Inside were folds of lead-like material – like the inside of a battery - into which were inserted a few, small, pellets of enriched uranium at regular intervals. While not totally innocuous - I needed to carefully wash my hands because uranium is toxic in the same way lead is toxic - I was amazed I could touch it and asked: *"Is this what all the fuss has been about?"* Most of us know so little about nuclear power generation that it has been easy to muffle our collective minds with misplaced apprehension.

Reactor vessels and fuelrods are designed to harness the natural radiation emitted from uranium and isolate, contain and capture the energy from these natural reactions in very small, almost impervious, spaces. A similar process is happening in our backyards all the time. Uranium is found all over the world in minute concentrations in soil. But there, the unstable, atomic-particle-ricocheting, chain-reactions, peter-out fairly quickly because there are lots of substances able to absorb the few neutrons uranium emits slowly: a cubic meter of soil or clay or rock can absorb a huge amount of radioactivity with little or none being emitted beyond a *metre* or two.

Further below the surface, towards the centre of Planet Earth, where the heaviest elements are found concentrated, the process does not peter-out as quickly and these sort of nuclear reactions, in Earth's molten metal core, create the release of energy that fuels our planet's volcanic activity and all of its tectonic plate movement.

To reiterate: man-made nuclear fuelrods are designed to contain and prolong the energy-releasing, neutron exchange processes. In the fuelrod, the unstable uranium isotope's radiant energy is deliberately directed towards both unstable and stable uranium atoms, and, as these absorb electrons and/or neutrons, they, in turn, become unstable and *"re-act"* or *"rebalance"*, by emitting more energy. At the high temperatures caused by this radioactivity, temperatures similar to those attained by other molten metal (*it generates heat as a fire does, hence we say a nuclear reactor "burns" metal*), the heat-excited atoms can even split into two: a process we call *"nuclear fission"*. In this process they *"fizzle"* into something else and two, or three or more, lighter elements can be created.

"Nuclear fusion" is a different process; it is the one that happens in our star, the Sun. In fusion, elements combine rather than split. The Sun's heat appears to be generated from collisions between the most simple sub-atomic-particle-elemental-arrangement known, the element hydrogen, combing with other hydrogen atoms. Fusion also happens in man-made fuelrods in a minor way when different elements combine or when an element absorbs enough stray particulate energy to cause the receptive atom to change *(transmute)* into a heavier element.

Modern reactor vessels contain and capture nearly all of this sub-atomic-particle-movement's heat *(energy)* to turn water into super-heated steam. Nuclear engineers use moderators *(surrounding water, graphite, helium gas, concrete or lead)* to slow-down fast-moving, sub-atomic particles and prevent their chances of escaping into space and burning *(contaminating)* people and materials outside of the containment structure. Moderators allow the minute, but very strong, atomic attraction and repulsion reaction forces of the elements' sub-atomic particles to assert their influence inside the fuelrod, and so, the moderators deflect particles of energy back into the neutron-absorbing fuel. Shielding-moderators facilitate, contain and prolong the slower, heat-producing, chain-reactions needed to manage a reactor's electricity-generating process.

Don't despair: that which you have just read is just about as complex as this nuclear power story gets.

We can create electricity because water *(or helium gas)* surrounding, or running past, the fuelrods captures heat and causes another body of water *(which is never-irradiated)* to become super-heated steam which, in turn, is used to turn turbines to generate electricity. As a technology, this heat-production process is now very well understood and when *"safety-first"* management procedures are diligently implemented, it has proved to be our most reliable method of generating clean, reliable, high-grade, electrical energy.

It is unreasonable and irrational to fear this technology: it is today the world's safest, power-generating, infrastructure. Nuclear power generation, carefully managed, has proved, and will continue to be proven to be a wonderful friend to all of humanity!

It is now important to address and confront the main ideas used to make the eminently manageable, nuclear-powered electricity production option unreasonably and irrationally feared by most of the world, but especially by Australians and New Zealanders.

The first lie

This is a distortion. We are told that *any form* of radiation is harmful to our bodies and will cause cancers and mutations. This is incorrect. More than being incorrect, it is irresponsible to propagate this fear-inducing myth.

We are surrounded by a sea of background radiation (*from the sun's cosmic rays, airborne dust and rocks, soil, sand, bricks and concrete*) at all times and our bodies depend upon some of these rays to manufacture melanin and vitamins. We also need to ingest radioactive substances, like the potassium found in bananas, to have our internal metabolic processes function correctly. Our bodies have evolved with mechanisms to adapt to slight variations in the intensity of these natural radiation exposures, and **in this fact, we find the key to the safe management of industries dependent on radiation emitting materials. Workers can be protected by ensuring that the intensity of exposure does not greatly**

exceed the exposure thresholds we are subjected to naturally.

Those who say that radiation of any sort or intensity is harmful rely largely on belief in a single, fallacious, principle called the *Linear No Threshold* (*LNT*) hypothesis. This idea suggests that all of the risks and dangers associated with high intensity, large, doses of radiation will continue to apply to even tiny doses, down to zero (*Members' discussions, ANA conferences 2006 & 2007, Australia's Uranium*, 2006, p277). This is very wrong. It is a concentrated, intense, dose - 10,000 times that which we would naturally receive on a daily basis - that could injure us to the point where our bodies will not recover. A less-intense dose, 1,000 times greater than the natural, daily, levels of exposure, will also cause some tissue damage, but it is likely that our body's natural defenses will expel the damaged cells, rejuvenate tissue and recuperate. Such intense doses are not probable in a nuclear power plant because, there, the current management of radioactive materials *limits maximum exposure to between five to ten times natural levels*, which, rather than being harmful, has been shown to have a "*tonic*" effect on human health. This "*tonic*" is the effect sought and found in natural spa baths and by our drinking natural mineral waters. Both of these activities expose participants and partakers to higher-than-normal doses of radiation without harm. The "*tonic*" effect is called *hormesis*: it stimulates our body to have the equivalent of an internal "*spring clean*". Also, slightly-higher-than usual, radiation-exposure-doses can be managed by the human body if they are substances that our bodies would normally expel, and this principle enables medical science to use irradiated material to trace blood flows, reveal clots, bone damage and other tissue abnormalities. The irradiated materials they use to make such analyses are those which will be disposed of by the body's natural waste elimination processes. Australian scientists are leading the world in many facets of this particular science, and in the production of medical isotopes, at its Lucas Height's nuclear facility.

To overcome contrived or irrational fear and manage radiation technology safely we need to understand the reality of the hazard more fully: **we fear that which we do not understand.** Bear in mind, that the creation of fear in its target audience is one of the strongest weapons in the propagandist's repertoire. It is known in circles of sophisticated professional communication that if a *"fear-inducing association"* is created with something unknown, even something that need not be feared, the emotional response will be particularly strong and it will likely occlude rationale thought. Nuclear-powered electricity production, rather than being associated with anything clean, useful, sustainable, profitable and manageable has been, misleadingly, associated with mutations, cancer and death. Contrived, fear-filled, associations about fission-powered electrical energy (*nuclear power*) inhibit rationale discussion and cause many people to remain close-minded and hostile to sensible applications of the technology. It is right to fear *and* loathe *and* despise nuclear weaponry, but nuclear electric power is proving a wonderful friend to all of humanity. It will provide our entire planet with the fresh water and the wealth to enable our world to become ecologically sustainable. Thus, it is very important to correctly understand *"safe"* and *"unsafe"* exposure levels and, especially understand, how the *"unsafe"* can be managed *"safely"*.

While low-level exposure registers negligible harm, prolonged high-level exposure must be avoided: it would be irresponsible and negligent of any organization, person or government to condone such doses or pretend that they cause no harm.

As a *"rule-of-thumb"* guide, a single, intense dose of radiant energy **equivalent to a person receiving the annual average *yearly* dose an adult could expect naturally, has**

no observable effect on health, but after this level, the first effects of radiation poisoning are discernable: nausea and vomiting. An intense dose four-and-a-half times the *annual average yearly* dose an adult receives naturally is called the *"LD-50" (lethal-dose, fifty).* It is expected that, in the absence of treatment, such a dose would result in a 50 percent chance of death (*Comby, 1994*). After sixty years of experience in monitoring the effects of radiation, the *International Commission on Radiological Protection* advises that annual doses below ten milli-Sieverts are not regarded as harmful and that the maximum permissible *dose* for any occupation should not exceed 20 milli-Sieverts over a five-year period (*total dosage over five years: 100 milli-Sieverts*), with a maximum of 50 milli-Sieverts in any one-year.

These recommendations are adhered to because there are two specific hazards associated with higher-levels of exposure to radioactive materials. They come from two forms of contact.

Intense penetrating radiation, that which comes from higher-than-natural-levels of energy impacting the body through neutron bombardment or energy rays in the form of alpha, beta or gamma radiation resulting in tissue being altered at the cellular level, or, in worst case scenarios, tissue being *"burnt"* or *"cooked"* (*as we would experience from being too long and close to, or wholly engulfed, by fire*); and,

By our internalizing irradiated particulate material, which, when ingested, inhaled or somehow absorbed into our bodies could expose internal tissue and organs to prolonged bombardment from damaging, ionizing, radiation.

The first hazard, *"penetrating radiant energy"*, that which can penetrate our skin and send radiant energy deep into tissue or organs, is hazardous when the dose intensity exceeds our body's natural defense capacity. It could cause our cells to malfunction chemically or it could damage one or both strands of a cell's DNA (*the material that programs how the cell functions*). But a natural defense process would then be activated. When a cell malfunctions chemically, it normally dies and is expelled from our body: this natural process is called "**aposteosis**".

If only one strand of DNA is altered by radiation, the cell can remain functioning properly because the duplicate, second strand, remains intact. The rare occurrence that results in both strands of DNA being damaged, either, causes the affected cell's death, or it can result in a cell reproducing without its proper function. This sort of false-cell growth can cause tumors or cancers, but a significant number of cells in a particular locale would need to have their functioning overwhelmed for a cancerous growth to commence. The far greater cause of such cancers, 98 percent plus, is chemical pollution, and the most prevalent, modern, carcinogen is not radiant energy, it is a chemical carcinogen, specifically, benzene, found in coal tar, petroleum, disinfectants, cleaning fluids and detergents (*Devra Davis, 2007*).

The chances of high-level radiation exposure danger can be made negligible in a nuclear power plant. The nuclear physicists' and nuclear engineers' solution to this problem has been to minimize the intensity, the duration, and the frequency, of exposure to any form of high-level radiation. To do so, they use the least possible amounts of radioactive material required for the work at hand; they minimize the amounts of these substances stored on site; they surround radiation-emitting sources with neutron-absorbing substances; they minimize and make secure the area required for containment; they clothe their scientists and other workers most carefully (*and make wearing radiation-detection and monitoring lapels mandatory*); drinking and eating is strictly prohibited in the laboratories; and all sites apply appropriate shielding (*screens that are largely impervious to radiation: mostly water, lead and concrete*). They also employ remote handling to deal with materials that may emit high levels of radiant energy and they monitor and restrict access to all activities.

The lists of regulations pertaining to helping create a safe work environment for nuclear power workers are huge, comprehensive, documents. Specially-trained officers are required to monitor, and exact, compliance.

The second hazard is materialized if we inhale, ingest or absorb irradiated, particulate material into our body and it becomes incorporated into our body's tissue. If it is a short-lived isotope, it will be a fast-emitter of ionizing particles and energy in the immediate vicinity of where it has lodged. This is dangerous because it can cause a *"nest of decay"* and this, in turn, can cause many cells in that *locale* to malfunction and die, or, it may result in dysfunctional, living cells and cause tumors or cancer to develop. **But, if it is a long-lived isotope, it will probably cause no harm because its rate of emission is so very slow** and the body's natural defenses will not be overwhelmed.

This is the case with uranium oxide and uranium ore: it is not particularly hazardous for its radiation emissions; rather, it is toxic if swallowed because it is a poisonous, heavy, metal. This is why we must wash our hands after handling uranium: the metal can be absorbed through our skin or we may ingest it by putting our hands to our mouths. If uranium is swallowed it can produce effects similar to lead poisoning. It is more hazardous underground and this is why uranium ore miners (*and handlers of uranium oxide*) wear air-filtering facemasks made of tissue.

But scientific enquiry, over the last hundred years, found that ***all miners***, not only uranium miners, were vulnerable to much higher levels of lung cancer than people who worked above ground. This is because one of the daughter decay products from the isotope uranium-235 is radon gas. In nature, radon gas is produced from naturally occurring uranium (*uranium is in all soils: it is a common element*). And wherever it occurs, small amounts of radon gas continually seep from the soil into the atmosphere. Radon is inhaled and incorporated into our lungs' tissue, regularly. Normal concentrations will do no harm, but if the gas is concentrated, as it can be in underground, confined spaces, it may decay (*transmute*) further. Being radioactive for only a short period, concentrated doses of radon can decay into more of the radioactive isotopes of bismuth and polonium than our bodies can efficiently, naturally, eliminate. These substances can become embedded in the lung tissue and their concentrated decay, in turn, may initiate cancer, but if underground miners and uranium handlers wear a simple paper filter for breathing, this gas is prevented from entering their lungs in harmful concentrations (*Williams, 1998, Cohen 1990, Keay, 2004*). Workers required to handle the uranium ore, at any stage, should always be clothed in fabric which is impervious to dust and prevents skin contact with the ore.

The second lie

The second position (*the uranium resource is finite and so nuclear power is not sustainable for energy generation*) is the most serious myth and the truth is so well disguised that it has taken several months of considered reading and thinking for the reality and the implications of the real situation to become apparent.

That uranium ore is finite and rare is true, but that it is, therefore, not a sustainable future fuel is a total fabrication. **Uranium will provide fuel for safe, high grade, base-load electricity production for the entire world for thousands of years**, even using current technology. Two of its natural, isotopic forms provide us with both the rare, unstable, isotope used to *commence* the chain of atomic reactions to generate energy, uranium-235, and the *"fertile"* material, receptive to capturing the emitted neutrons, stable uranium-238. When combined, contained and controlled, this process produces heat, which, in turn, creates other unstable, or fissionable, isotopes needed to sustain energy production in a reactor vessel. The reactions *"peter-out"* because they eventually result in the production of stable isotopes. When this happens, a reactor's fuelrods are changed. The reactions also produce, as a by-product, several plutonium isotopes. Plutonium, a reactor-made element, is formed when uranium absorbs more neutrons, so it is an even heavier atom than uranium, and one of its isotopes, in particular, the fissile *(unstable)* isotope, plutonium-239, has the same wobbly properties as the rare unstable uranium-235 isotope. When combined with uranium-238, plutonium can also be used similarly, to start the chain-of-reactions to generate heat suitable for electricity production. This particular reaction is the basis of the new, mixed-oxide fuel reactors *(sometimes called Generation IV reactors)*.

Fuels made with a mixture of man-made plutonium allow every skerrick of *stable* uranium oxide to be used to produce energy. It extends the electricity-generating potential of the uranium ore resource for thousands and thousands of years.

The easily accessed, scratch-off-the-ground, uranium-bearing, ore is indeed finite. It is estimated, in total, globally, to be a seventy to one-hundred year maximum, mine-able commodity. And this *"life of mining"* was allowed to be thought of as the *"life of the ore body"*, hence, uranium became widely regarded as a limited and unsustainable resource. But, uranium-based, nuclear-powered electricity production is eminently sustainable and the ore must, therefore, be almost priceless. This potential, and its scarcity, are the main reasons for obfuscating the truth about uranium's sustainability *(see the Australian edition of this book for important supplementary material provided by Professor Leslie Kemeny, one of Australia's most distinguished scientists: see bookshop at www.digitalprintaustralia.com.au)*.

Uranium's concentrated appearance in an ore body is rare and precious. Uranium ore is the end-product of millions, if not billions, of years of erosion: the result of the weathering of ancient, ferrous (*iron-ore-bearing*), mountains. Being the heaviest element, it was deposited, and concentrated, where the water flows slowed: in ancient riverbeds and pondage sites (*like Olympic Dam and MacArthur River*). The mountains have long since gone, but the concentrated ore remains alongside other heavy metals, such as lead, gold, zinc and silver. In our collective ignorance, Australians, Africans and several other nations have allowed some of the world's richest ore bodies to be entirely owned and mined by a very small number of beneficial nations. We, the countries whose heritage has been "*exploited*", are entitled to feel aggrieved. However, that exploitative time is now rapidly ending and in its place we have the principle of mining royalties in future being paid in the form of an ore (*mined product*) apportionment (*thirty percent or more*). Where this form of resource sharing is no longer possible, because the ore bodies are exhausted, retrospective recompense through opportunities for personnel training and nuclear powered electricity infrastructure development aid, must now be seriously instituted. Proposals like these, and other new forms of more-equitable resource distribution, have the potential to rapidly help alleviate poverty and help the idea of globalization to mature into a more morally acceptable form of international trade.

The third lie

We have been led to believe that only 0.7 percent of the uranium ore body (*the unstable isotope, uranium-235*) is useful for fuel production and that depleted uranium oxide (*uranium-238, the 99.3 percent left over when the lighter, uranium-235, isotope is removed*) from the "*yellow cake*", is a waste product. This is hugely misleading. The word "*depleted*" implies that only the enriched uranium oxide (*that containing high concentrations of the isotope uranium-235*) is useful for power generation. The rest, "*depleted uranium*", is then regarded as a "*waste*" and nuclear waste of any sort is thought to be harmful by the public at large.

The truth is that *all* of this uranium, enriched and depleted, can be used to generate electrical energy. And, while supplies of the natural fissile isotope uranium-235, only 0.7 percent of the ore body, could be quickly exhausted, recombining *depleted ore* with a similarly functioning isotope of plutonium, *will* provide power for the whole world for thousands of years.

I digress to provide a little more background information. Uranium has several isotopic forms. To be useful as a fuel, the lighter isotope, the unstable, "*fissile*" uranium-235 (*0.7 percent of uranium oxide*) is mechanically enriched to a concentration rarely found in nature, to three percent (*hence "enriched uranium"*). This enriched material is separated from the bulk of the uranium oxide, predominantly uranium-238 known as "*depleted uranium*" and pelletized. Following a very particular procedure, it is used to start the chain of reactions that result in energy release. In this process it can "*fizz*" and change into other, lighter elements. These new elements can "*poison*", or prevent the chain reaction and, because of this, they are the only genuine waste in a fuelrod, about three percent of its used-content. Even so, this waste can now be disposed of safely. All the rest of the metals can be recycled to generate more power.

In nature, it is very rare for enough of the isotope uranium-235 to concentrate but it has done so millions of years ago and several *natural* nuclear reactions occurred in an ancient riverbed in Africa at a place called Oklo. By studying these sites, scientists have learnt how to safely dispose of genuine nuclear waste. The Australian invention "*synroc*" is part of the fruit of this enquiry. It is material made from copying the type of mica-rich sands that helped to isolate and contain these ancient, natural, nuclear reactions. Synroc encasement, coupled with underground, deep geological burial, is now the most effective, permanent method of containing radiation-emitting substances.

A fuelrod works like a giant battery, but instead of producing a direct current, in an electricity-producing reactor, the fuelrod generates heat in water engineered to flow past it, in a sealed, closed system. This hot water heats another, separate, body of water so that it becomes super-heated steam and, under pressure, this steam is channeled to turn turbines to produce a regular, alternating, current. Alternatively, helium gas can be used to transfer the heat to produce super-heated steam. In both types of (*pressured water and gas-cooled*) reactor, the turbine-turning steam **never** comes into contact with irradiated material, so irradiation opportunities are minimized.

To reiterate and elaborate: uranium is extracted from an ore body and processed to make uranium oxide, "*yellow cake*", at the mines. Enriched uranium is that which is derived from more processing designed to concentrate the lighter isotope, uranium-235. It constitutes about three percent of a fuelrod. All of the remaining uranium is called "*depleted uranium*" and has been misnamed "*waste*". But while uranium-235 is needed to start the chain-of-reactions that result in heat being generated in the fuelrod, a great deal of the energy produced is actually a result of it being placed beside the heavier, neutron-absorbing isotope, uranium-238. When the fissile (*unstable*), uranium-235, emits neutrons (*a natural process, causing radiant heat*), uranium-238 accepts one or two of them (*fusing and releasing more energy as it does so*), thus changing its mass, or it can manifest as plutonium, which, in turn, can degrade back into the more stable uranium-238, releasing more energy as it does. The inherent instability of these new, heavy, reactor-made isotopes causes them to emit neutrons, many or just a couple, producing more *fission* energy in the process. These processes of fusion and

26

fission provide us with all of the energy produced within the fuelrod. Even when these reactions have petered-out, some unstable uranium-235 (one percent) and plutonium (one or two percent) remain in the "*spent*" fuelrod. This remnant can be extracted through reprocessing and, by mixing it with "*depleted*" uranium oxide, uranium-238, these unstable isotopes can be re-used to "*kick start*" a new chain-of-reactions long after the original supply of the rare, mined-uranium-235 is exhausted.

This next part of the story is significantly simplified, but any simple, first year university-level chemistry or physics book can provide you with more precise knowledge. When uranium acquires another proton in its nucleus (*extra neutrons can change, "degrade", into a proton*) it can then possess enough distinctly different and stability-holding characteristics to warrant it being called a new, man-made element (*elements are actually determined by the number of protons present in the nucleus*). The heavier uranium isotopes transmute into the new element plutonium and several other heavier elements called "*transuranics*".

Like uranium, plutonium has several isotopes, but the one most useful in producing electricity is the one that emits neutrons at a rate similar to uranium-235: plutonium-239. Plutonium-239 is "*fissile*" in the same way uranium-235 is fissile. When enough plutonium-239 is concentrated (*and it can be mixed with uranium-235 to create a mixed-oxide fuel*), then placed alongside the uranium that is receptive to accepting neutrons (*uranium-238*), it will also generate energy that is easily captured for electricity production in a reactor vessel.

This neutron emission process is **natural**. It is representative of the natural, inevitable, change in the internal structure of certain "*unstable*" atoms. The internal, very dense, nucleus of an atom is bound by very strong forces (*remember: they are the forces that hold the material world together*) and when these bonds are disturbed, a great deal *more* energy is released than is released when atoms combine at their outer extremities (*through electron exchange*), such as happens when wood, predominantly carbon, is burnt (*combines with oxygen*) to produce carbon dioxide.

To summarize: the resource, uranium-235 is naturally *very* scarce and rarely able to concentrate. When we isolate it and concentrate it in small amounts, and then cause it to interact with uranium-238 in a carefully controlled environment, lots of energy can be generated from very small quantities of uranium-238. A small amount of plutonium is manufactured as a by-product. Using plutonium in the same way uranium-235 is used, allowing it to *react* with *depleted* uranium-238, will enable the entire uranium oxide ore body to be used to generate electricity for thousands of years.

Thus, it can be seen that uranium-238, "*depleted uranium*", is not a waste at all but a critical component of both the enriched uranium- and plutonium-dependent, nuclear reactions that can produce heat for electricity generation.

The other dimension to this is that spent fuel is also not *waste*. A spent fuelrod contains one percent enriched uranium, one percent plutonium, ninety-five percent *depleted* uranium and three percent genuine irradiated waste (*called the light-weight fission products or actinides*). All of the remnant material, except the light-weight fission products, can be recycled as fuel for the future *using existing* technology. All of the uranium and plutonium in a spent fuelrod, ninety seven percent of the content, can be recycled. The three percent that is genuine waste (*the lighter fission products*) can be either burnt to produce more electricity or safely disposed of in synroc (*from which they cannot leach*). (*Down the track, it will be burnt in reactor vessels able to operate at higher temperature – but providing this expensive, extra, infrastructure is thought to be too costly at this stage.*) "Synthetic rock" (synroc), and even some radioactive materials, can be buried deeply and backfilled with salt or clay so that they will then pose *no possibility of harm* to the living biosphere: but even that treatment for synroc will be *"overkill"*. Synroc emits less radiation than does most of the igneous rock above which we build our homes.

So, the 97 percent of the residual material contained in a spent fuelrod provides us with the raw material for more energy production: this is why **spent fuel is carefully stored all over the world**. It is wrong to think that spent fuelrods are waste. For example, by simply reprocessing and utilizing its stored *"waste fuelrods"* to harvest plutonium and then mixing this with depleted uranium oxide, the United States will be able to power its existing nuclear power plants with mixed oxide fuel for at least 4,000 years – it need buy no more uranium from Australia (i.e. *if the US has been generating nuclear power for 45 years and 95 percent of the energy generating potential remains in their stored spent fuelrods, they have, upon the first re-cycle, 45 x 95 = 4275 years of electrical energy in storage*).

Recently, a new generation of gas-cooled reactors, Generation IV, has been specifically designed to use the reprocessed fuel (*a mixture of plutonium, enriched and depleted uranium*) hence, Generation IV reactors are also called *"mixed-oxide reactors"*. But older, light-water, pressure reactors can also be run using *"mixed-oxide"* fuel. High temperature, fast neutron reactors, or *"breeder"* reactors (*so called because they produce more plutonium than they consume*), are those that are designed to produce more than one percent plutonium for harvesting and applying as a mixed-oxide fuel; but the world will only need very few of these to produce mixed oxide fuels economically. Globally, we already have enough plutonium to enable electricity production from uranium oxide for many hundreds, if not thousands, of years.

To conclude: the idea of uranium being a finite fuel resource is wrong. Fuel for future nuclear electricity production, long after all the uranium has been mined, can be extracted from spent fuelrods or manufactured by using plutonium mixed with *depleted* uranium as the receptive or *"fertile"* material needed to prolong power production, even using current generations of reactor engineering technology.

Nuclear-advanced nations, until very recently, chose not to recycle fuelrods and use the more economical and resource efficient, mixed-oxide fuels. Instead, they have maximized the resource-gobbling, enriched, uranium-only fuel processes and this dependence helped facilitate their artificial need to continue importing uranium oxide. It has been in their national interests, not to use recycled fuel while they have had unfettered access to, inexpensively, acquiring uranium ore from various locations all around the globe and, by so doing, they have been able to manufacture vast amounts of uranium oxide and store it for future use. France has 100,000 tonnes of uranium oxide in store and in this, they have fuel for themselves, and all of the European nations they currently supply electricity to, for 5,000 years (*ANA newsletter, 2009*). The United States, Britain, Russia, and Canada, now also have enough uranium and plutonium to provide them with electrical energy security for thousands of years.

But besides generating amazing energy security for the few, knowledgeable nations, there has been another, hidden, consideration at play as an undercurrent, socio-economic, force propelling the manufacture of misinformation to disguise the truth about nuclear power.

Utilizing mixed oxide fuels will enable electricity to be generated at a fraction of its current cost. This cost-effectiveness alone would have made all other electricity generating utilities, especially those dependent on fossil fuels, **immediately** economically redundant. It is to humanity's shame that the economic considerations propping-up outmoded fossil-fuel-burning technology have usurped, for forty years, the plea from conservationists and concerned scientists for the world to cease polluting and befouling Earth's atmosphere. One can only hope that more enlightened policies will now, very carefully, manage the world's remaining ore to provide prosperity and energy security for its entire global populace. To this end, our planet must rapidly adopt nuclear power. It is especially needed because gas and coal-burning for electricity continues to create havoc in our atmosphere and we have much better uses for our coal: uses that will enable a great deal more sustainable employment and wealth to be generated than burning the resources allow.

The fourth lie

The fourth position is a totally out-dated, fear-driven myth: nuclear power is *so* dangerous that it cannot be managed. Absolute nonsense! Sure it is dangerous. It is dangerous in the way any source of intense heat is dangerous and it would be particularly stupid to approach utilizing it without correct safety precautions. But these safety precautions exist and have done so effectively for more than forty years. Nuclear power production is a mature technology (*nearly seventy years old*) and it has evolved to be, without doubt, the safest (*and cleanest*) power-generating infrastructure in the world. Here are just some of the comparative facts as provided in the compendium *Australia's Uranium*, 2006:

Between 10,000 and 15,000 coal miners are killed around the world each year;
Coal-fired power stations have killed 6,500 people since 1997;
Hydropower production has killed more than 4,000 (*mainly through dam bursts*).
The Bhopal chemical accident resulted in 4,000 immediate deaths, with another 15,000 people killed within two years.

Contrast these significant losses of life with the comparable few fatalities associated with the Chernobyl accident where 32 fire-fighters were killed (*their lives may have been saved had they worn the protective clothing mandatory in other developed nations' facilities*) and another 134 emergency workers were diagnosed with acute radiation sickness (*47 of these men dying over the period 1986 – 2004*). The ensuing, operable, thyroid tumors affected 4,000 adolescents and have resulted, to date, in nine deaths: a tragedy that may have been entirely obviated had the children been provided with an iodine supplement (*Australia's Uranium, 2006, pp322 - 333*).

While I do not want to belittle the seriousness of the Chernobyl accident (*it is the only accident at a commercial nuclear power facility that has resulted in sections of the general public receiving more radiation exposure than they would from natural sources*), **it was totally avoidable** and twenty years on, we can see that it was an accident that forever changed reactor design and the culture of nuclear science. It is a fact that there have been **zero deaths** associated with civil nuclear power plant infrastructure since the Chernobyl incident. **No other power-generating utilities have a comparable record.**

Post-Chernobyl, the world of nuclear scientists and engineers combined to share innovations, ideas, technology, and management procedures, so that such eminently preventable accidents would never happen again. The World Association of Nuclear Operators, WANO, was founded after the Chernobyl accident to ensure that nuclear power production is managed with strict attention to "*safety-first*" principles. Since its formation, there has not been one single incident of public harm from any licensable reactor in the Western World. This amazing safety record has been achieved through active, co-operative, international networking to share operational, training, management and technical innovations "*par excellence*". Can good come from bad? Yes: we have learnt to use nuclear technology more carefully and safely as a result of the two preventable accidents at civil nuclear electricity production plants: Chernobyl and Three Mile Island (*described later, in the "Sixth lie"*).

Compare this extraordinarily rare incidence of accidents and death in the nuclear power industry with the high-risk of death associated with using an automobile. For decades, several hundred, sometimes thousands, of Australians have been killed each year; in the US, as many as 50,000 are killed annually; and, globally, 1.2 million people die each year in car accidents (*these and other statistics are found in Australia's Uranium, 2006, and William's, 1998, text: What is Safe? The Risks of Living in a Nuclear Age*). Simply "*sniffing petrol*" can cause instant brain damage and inhalation will result in death in confined spaces. It makes more sense to ban cars and petrol if we are concerned with elevating public safety than it does to ban nuclear electricity production.

Under the licensing regulations and international agreements that now prevail, average radiation exposure for nuclear fuel industry workers is controlled and reduced to a maximum of approximately two to three times the daily natural radiation dose all humans receive from sunlight, the soil we stand upon and the materials used in constructing the buildings we shelter under (*cement, granite, bricks*). We would die without this natural, surrounding sea of radiation: it represents and maintains the conditions under which we have evolved.

Continuous, natural, background radiation exposure varies between 2 to 4 mSv (*milli-Sieverts*) a year. Most nuclear reactor workers at the Australian facility, Lucas Heights, receive significantly smaller doses than the international allowable annual dose of 20 mSv. For example, the *average* exposure for those working at Lucas Heights is an extra 1 mSv a year. The extra, limited, exposure nuclear industry workers receive has not been shown to be injurious, in fact, low level radiation exposure, over a long period of time, can have a tonic effect on our health: a phenomenon called **hormesis**. It is high level, intense doses, greater than ten thousand times the natural daily exposure level, which maim people to the extent that they will not recover. These doses have very rarely happened, but there have been one or two instances recorded when medical radiation technology has failed to correctly calibrate, or regulate, the dose of administered radiation and some patients have been severely burnt *(even, horrifically, cooked)*. I, personally, believe that these medical-radiation-accidents are the significant reason explaining why many medical practitioners refuse to support a nuclear powered electricity program for Australia. But nuclear power facilities are not designed to deliberately target humans with concentrated radiation *(such as are medical x-ray and magnetic resonance imaging machines)*: they are deliberately designed to prevent any possible excessive exposure *(see Williams, 1998; Australian's Uranium, 2006, Chapter six; Cohen, 1990)*. Visiting Lucas Heights is not particularly dangerous: it is far more dangerous to have any form of medical or dental x-ray, yet these far more substantial risks are happily embraced as routine procedure by most of the developed world's population.

How does the scientific community prevent excessive exposure? The precautionary principles used by the developed, nuclear-savvy world to protect nuclear power plant workers and the general public from intense exposure are based on trenchant licensing standards that:

Limit exposure time;
Use effective shielding *(through concrete, water and lead)*;
Enforce the maintenance of safe distances *(radiation intensity diminishes with distance)*;
Implement passive safety features *(automatic measures that do not require humans to operate, activated through pressure*

and temperature differentials, gravity, high temperature resistant materials, and natural convection), i.e. many reactors **will passively shut down instantly** if an anomaly were to happen;

Contain waste (*keeping material in the smallest possible space and out of contact with biological material until reprocessing or deep geological burial can happen*);

Minimize dissipation (*ensure dispersal of irradiated materials does not exceed the natural levels of radiation in the environment*);

Have *at least* two or three back-up levels for ensuring containment, rapid shutdown and ensuring the alternative water supplies are always available;

Monitor radiation exposure continuously;

Conduct regular, independent inspections with a view to enforcing compliance with international legal codes and conventions, and so forth (sources: ANSTO, various educational pamphlets; *Australia's Uranium, 2006*, Cohen, 1990; Keay, 2004); and,

Finally, all modern installations rely upon "*proactive*" safety-first engineering and design principles, rather than "*reactive*" attention to safety issues. This means that every imaginable type of accident and malfunction is considered and risk factors are eliminated from, or compensated for, in a utility's design, construction and operation.

If these precautions, coupled with the most sophisticated engineering and mechanical designs, were not effective the United States, Germany, Switzerland, France and Britain would not be re-licensing their thirty to forty-year-old reactors for another twenty years of operation. Nor would Argentina, Armenia, Belgium, Brazil, Bulgaria, Canada, China, Czech Republic, Egypt, Finland, Hungary, India, Indonesia, Iran, Israel, Italy, Japan, Kazakhstan, Malaysia, North and South Korea, Lithuania, Mexico, Netherlands, Pakistan, Poland, Romania, Russia, Slovakia, Slovenia, South Africa, Spain, Sweden, Taiwan,

Turkey, Ukraine, United Kingdom, Vietnam (*and the list increases annually*), be commencing or expanding their nuclear power production facilities, adding to the more than 440 reactors (*30-odd under construction this year, 70 more planned for construction in the immediate future*) already *safely* operating around the world today.

The fifth lie

This position represents a particularly unpalatable political possibility. It is not an end-use certitude. It proposes that because uranium enrichment results in plutonium, and plutonium, like uranium-235, can be a feedstock for nuclear weapons: nuclear weapons production is, therefore and forever thereafter, a corollary of plutonium production.

This conclusive suggestion is the result of the most superficial level of analysis possible. And so I respond at the most simple level. Scientists and terrorists can do anything they want with enough technical know-how and resources, but to use plutonium in this manner is incredibly wasteful, phenomenally expensive and hazardous to those who would develop weaponry: it is likely they would be killed in the process. This is because one of the several plutonium isotopes produced in a reactor emits neutrons far more rapidly than the others: it has the potential to harm its handlers and cause a bomb to prematurely detonate or fizzle.

Separating these isotopes is not a requirement for using plutonium to generate power – it would be for bomb production.

Such separation, and weapons end-use, requires the application of a science that represents technical and conceptual complexity beyond the reach of almost all of the world's population.

Furthermore, any nation that chose to turn plutonium into weaponry is squandering its amazing potential for safe, sustainable, electricity production. It is simply too wasteful of resources and money to do so: it is far too valuable for energy production to use in this fashion. But, to minimize the risk of misuse, it is important that the world does not produce a surplus of this material and that it be manufactured on a *"meet needs basis only"* in the future by authorities who can be trusted not to direct any to weapons manufacturing.

As an option, the potential for weaponry development is being made redundant by the concept of *"fuelrod leasing"*. This option ensures that in future, fuel will be reliably provided by responsible nations, those which are demonstrably committed to not developing, or using, *"first strike"* nuclear weaponry, such as is Australia.

By leasing new, and reprocessed, fuelrods, less stable and less peaceful, developing nations can be provided with access to enabling electric energy and so have the opportunity to achieve levels of prosperity that enable them to become more politically stable. This provision will help facilitate their becoming ecological sustainability: a critical aspiration for the entire world. When peace-loving nations assume responsibility for manufacturing and providing a secure, leased, fuel supply, under existing IAEA ore-supply restrictions, the possibility of the fuel-supplied nations diverting plutonium into weaponry, will be minimized, not exacerbated.

To help you to put the dangers of directing uranium or plutonium to weapons manufacturing into perspective: there are two opportunities for weapons manufacture using these substances specifically. One stems from the enrichment of uranium to produce uranium-235 to concentrations far greater than reactor fuel requires (*ninety-five percent plus purity*); the other stems from manufacturing pure plutonium, either in the spent fuel recycling process or through manufacturing plutonium in *"fast breeder reactors"* then concentrating and separating its fissile isotopes – an incredibly costly, technically complex and hazard-filled task. We are fortunate that uranium is a rare, closely-monitored resource because this makes tracing its movements around the world far easier. The strictest diligence must always be applied to this administrative, and security, function: we have not always succeeded and must now collectively strive to make this security infallible (*see "Security breaches at world's nuclear sites" Nuclear Power Daily, 28th April, 2010*).

Fuelrods assembled for power production require only a very slight increase in the concentration of the unstable isotope uranium-235, up to three or, at the very most, five percent (*the concentration process is called enrichment*). To separate and concentrate this uranium isotope, through further enrichment processing, to produce weapons-grade ore (*95 percent concentrations plus*) requires a prohibitively huge amount of uranium and very sophisticated equipment (*an amount equivalent to one entire year's production from an established Australian mine*). To date, the prevailing, rigorous, international monitoring of uranium use has foiled any such illicit attempts.

To extract plutonium from spent fuel-rods requires sophisticated, remote handling and expensive chemical or laser treatment before it can be "*recycled*" as mixed oxide fuel. It is material only able to be manufactured in a reactor and normally represents less than one percent of fuelrods' by-products. So, to manufacture plutonium for bomb-making, a terrorist needs access to a nuclear power plant to procure the by-products of the conventional reactions (*used fuelrods: which can only be handled remotely for a number of years*) **and** a reprocessing plant. Those facilities amount to billions of dollars of investment and scientific know-how. It simply does not make economic or material sense to go to the phenomenally expensive and technologically-complex lengths required to process the material to bomb-grade concentrations.

For Australians, now considering supplying fuel to our region through fuel-rod leasing and re-processing, the very good news is that plutonium can be simply and permanently contaminated with neptunium during a prolonged recycling process. This contamination ensures that **no** explosive potential remains and decontamination thereafter is simply not practicable nor is it feasible economically. And when an internationally-trusted country, like Australia, commences to enrich uranium to manufacture fuelrods to produce electricity for nearby neighboring countries, **and** assumes responsibility for recycling them (*thus removing the possibility of weapons production*), I believe our entire world will become safer and more prosperous.

The alternative to uranium/plutonium energy-based production is the thorium fuel cycle which produces no plutonium, but globally, like uranium, stocks of thorium are also extremely limited: it is a far more scarce resource than even uranium (*see Australia's Uranium, 2006, p 89 & 666*).

There are dangers associated with handling spent fuel. It does emit radiation and remains "*hot*" for a number of years, but the prevailing scientific opinion is that, even without recycling, its extreme danger to the world at large will be so greatly diminished after ninety years that it could be disposed of in deep geological burial safely. But we are able to recycle the spent fuel much earlier. It is stored in water (*a great neutron absorber*) until remote handling allows us to separate the metal and its useful isotopes. This material then becomes the basis of mixed oxide fuel and using this fuel will reduce the cost of electricity production thereafter, perhaps by as much as ninety-fold.

To conclude, it is misguided to overly fear and assume that weapons production is an automatic corollary of plutonium production for the following reasons:

The equipment and resources required to produce 95 percent plus percent concentrations of *pure* plutonium-239 or *pure* uranium-235 are beyond the budget and technical capacity of most nations, let alone disaffected terrorists;

Bomb production requires implosion and explosion that, in turn, is reliant on split-second timing to coincide with neutron emission (*which must be precisely calculated*) and unless it is conducted under the most stringent conditions, the substances' inherent instability will probably result in either, the bomb fizzing - failing and contaminating the would-be bomb builders with short-lived radiation – or, the bomb could prematurely detonate and very probably cause the death of those who attempt to manufacture such weapons.

Terrorists have always had access to simpler, cheaper, more certain and effective, means of disruption and harm;

The international community is committed to maintaining **the best possible security, high-level military surveillance and tracking for fuel manufactures and users.** It is up to us, the world's citizenry, to make sure that this activity continues and improves;

Considerately and generously managed, the worldwide adoption of nuclear power (*able to produce electrical energy at much less than the cost of current electricity*) will aid global prosperity. It is tremendously important to appreciate that if prosperity, based on the global sharing of nuclear powered electricity technology, were realized, there would be a significant reduction in the reasons for terrorism to manifest. Economic prosperity is a hugely stabilizing quality. The ideal of instituting mechanisms to global share this resource is an ideal we must now strive to uphold and institute.

The sixth lie

To believe this sixth position is the result of a misconception. We are told that a nuclear power plant could explode like a bomb and that the Chernobyl hazard remains in all nuclear power facilities. This is simply wrong.

Firstly, nuclear-powered electricity plants cannot explode like bombs: no explosive or detonating materials are part of the engineering requirements of a nuclear power reactor. Nuclear reactors operate like giant batteries and use fuelrods to generate heat. Heat production is controlled and transferred to turn water into steam. A reactor's energy is captured and contained, not dispersed and wasted as it would be in a bomb.

Remember: spontaneous changes to the nuclei of atoms release a great deal of energy. The heat-generating process, resulting from the chain-of-reactions caused by excess neutrons from the unstable uranium-235 or plutonium-239 bombarding, and being absorbed by, stable uranium isotopes, can be precisely **controlled.** It is controlled by the way the fuelrod is configured (*using a very little amount of unstable uranium*) and through the use of moderators and the insertion of neutron-absorbing material. Moderators, such as water and graphite, surround the fuel assemblies and slow neutron movement so that their activity is contained within the fuelrods. They prevent stray neutrons from escaping from the reactor vessel. Furthermore, all modern reactors are designed so that the control rods, made from neutron absorbing boron and cadmium, can be dropped into the reactor core, between the fuelrods, to **instantly stop the reaction**. In IAEA approved reactors, the control rods are designed to operate automatically if there was an earthquake, if there was a total failure of electricity or water, or if, for any reason, all control systems malfunctioned or were destroyed.

Secondly, reactors like the Chernobyl reactor would never be licensed today *anywhere* in the world. The Chernobyl reactors represent an archaic design, which, when marketed to the developed world more than thirty years ago, was rejected because of its lack of safety features, particularly the lack of a containment vessel (*Hardy, pers. comm., 2007*). All of these archaic reactors are gradually being closed as more efficient and safer models are commissioned. The remaining few in use have been upgraded and made safer since the tragic Chernobyl incident blighted nuclear electricity production worldwide.

In the entire fifty-year-plus history of civil nuclear electricity production and among the 440 plants-plus now operating safely around the world, there have been only two serious accidents: Three Mile Island and Chernobyl. The real tragedy of these unfortunate accidents was not the radiation they emitted, but that they were **totally preventable.** They each happened through inexcusable combinations of mismanagement, which, with hindsight, I find, incredulously, exposes each community of scientists and operators to the suggestion that they are culpable. Before the Three Mile Island incident, no-one knew what really would happen after certain temperature thresholds were exceeded (*on reactor vessel structure and containment materials*), nor could they predict with absolute certainty the new combinations of atomic matter likely to form, nor how rapidly neutron-absorbing control rods would take to slow and completely stop the reactions. These questions were able to be answered after the Three Mile incident. It is a cruel fact, both of these "*accidents*" gave a great deal of knowledge to both of the, then conflicting, sides of the *Cold War* divide. But from these terrifying mishaps - travesties of management - the global nuclear science community was provided with the unifying impetus to implement far more stringent controls, similar to those we use to guide air traffic to safely take-off and land. The nuclear power installations that are approved by, and come under the surveillance and regulatory auspices of, the International Atomic Energy Agency, now operate with thoroughly trained operators, men (*and women*) who exercise precise micro-management regimes and are supported by multi-level, fail-proof, automatic, pro-active not reactive, safety-first, control mechanisms.

It is important to understand the actual causes of the two nuclear power accidents to help dispel this particularly alarming, *"explosive"* myth. A nuclear power-generating plant will not, and cannot, explode like an atomic bomb.

In the Three Mile Island reactor, a partial *"melt-down"* occurred. The *"melt down"* is reported to have been caused by the failure of the local water authorities to advise the plant operators that their (*and the entire surrounding area's*) water supply would be stopped for a couple of hours. This meant that that reactor's in-flowing water's requisite cooling mechanism, a function of it flowing past the fuelrods, failed (*it was not then mandatory to have several "back-up" alternative supplies of water if the reactor's design depends on water for cooling*). Inside the reactor, a pressure value remained open in response to the buildup of pressure caused by the water inside turning to steam. From this opening, some irradiated steam escaped. Thanks to sound engineering, the structural and design integrity of the plant contained the damage resulting from the system temporarily overheating and prevented the possibility of significant outside environmental contamination. Contrary to public belief, the accident caused **no** deaths and the exposure from the escape of a *"slight wisp"* of radioactive xenon gas only manifest in the surrounding five kilometers before dispersing to normal background levels. As such, it was equivalent to three times that of natural exposure. Medically, this level of excessive exposure is not thought to be dangerous to the health of living creatures. The prevailing atmospheric conditions caused the gas to disperse fairly evenly and prevented the possibility of individual contamination through a single, extra-large, dose. In the opinion of the experts who assessed the accident, it was, fortunately, a fairly harmless incident (*find a more detailed description in Australia's Uranium, 2006*). Through the detailed, post-event, analysis, a great deal more was learnt about how to build and operate nuclear installations with, what now amounts to, unparalleled safety.

At Chernobyl, a melt-down happened and it was not contained. It happened in one of four *"archaic"*, Generation I, reactors, designed and built without a containment structure and incorrectly experimented-upon by poorly informed, poorly trained, operators. It was a dreadful incident. It resulted in the direct deaths of thirty-one fire-fighters and subsequently, thirty more *"clean-up"* workers. The experiment, according to my readings, was to reduce the water flow to slow the turbines to test at what lower rate-of-flow would allow the turbines to turn and still maintain stable electrical energy output. But the operators, electrical engineers not nuclear scientists - people not familiar with the reactor's design and heat-production processes - did not foresee that in slowing the water flow rate, they also reduced its cooling function. The slow-moving water boiled in the reactor, then turned into steam and steam has no cooling properties. Thus, the reactor quickly overheated and caused an explosion of *"steam"*, like a blocked pressure-cooker, and blew the inadequate lid off the reactor vessel. The damaged reactor, surrounded by graphite and very high molten metal temperatures, then caught on fire and caused a second steam explosion, tipping the reactor vessel over, spilling its contents and preventing the insertion of the neutron absorption rods. It was this second steam explosion and the fire and its smoke that carried irradiated particles into the atmosphere and created an incident where the public in the immediate vicinity were exposed to significantly more irradiated material than natural conditions normally allow.

The published scientific consensus about this event is that we will not know for many years what the actual effect on overall civilian health might be. Those in a position to make this assessment (specifically, *The Chernobyl Forum* as reported in *Australia's Uranium, 2006*) believe that it might result in a three to four percent increase in the established rate of cancer for the

600,000 residents in the most contaminated areas, and 0.6 percent increase for the whole region. The dose received throughout the rest of Europe was one-fifth of normal background radiation and, therefore, was thought not to be intense enough to be particularly hazardous (*Williams, 1998*). It is also thought that had the firefighters at Chernobyl worn adequate, protective clothing, like that which is mandatory in the Western developed world, it is probable they would *all* still be alive today.

To place the accident in a different context altogether, the three adjoining sister reactors at Chernobyl continued to operate, fully-manned, and still do to this day (*though one is in the process of being dismantled now*), albeit with continuous, necessary, substantial improvements to provide more secure containment, better site management and overall safety.

Since the Chernobyl tragedy, the international nuclear physics and engineering community made a dedicated, **huge** effort to **never** allow such incidents to happen again. The World Association of Nuclear Operators, WANO, was then formed and, subsequently, it has studied every single technical anomaly and managerial mishap in every one of its associated plants to determine the causes of the problem; it then makes a concentrated effort to overcome such situations, permanently. (*The writings of Bernard Cohen, David Williams and the regulations pertaining to the International Atomic Energy Agency's activity, some of which are accessible via the web, provide a thorough introduction to these internationally approved, and monitored, safety-first measures.*)

Consequently, the global nuclear science community calls nuclear power for electricity production a "*mature technology*".

Today, **without contest**, nuclear reactors are the world's safest electric power production plants. That they are able to be managed safely is to the absolute credit of the many thousands of tireless nuclear community workers who have dedicated their lives to making this so.

The seventh lie

Waste is not manageable. This is a total myth. Serious propaganda has been employed here because there is almost negligible *real* waste in the entire industry. It is very important that the global community commence to respect the knowledge yielded from careful scientific enquiry on this topic. This knowledge confirms that we are able to manage spent fuel safely. If we remain ignorant, it is likely we will remain fearful and susceptible to misinformation and continue to make serious, incredibly costly and harmful-to-the-biosphere, mistakes in our management decisions.

Nearly all the material that has been misnamed as "*waste*" (*that found in "used" fuelrods and the material, "depleted uranium", left over from uranium enrichment after the rare isotope, uranium-235 is concentrated and removed*), when reprocessed, will be able to be used as a future fuel. So it is not waste: it is fuel for the future. The management of reactor-exited material then becomes, in part, a "*how to store used-fuelrods safely until they can be reprocessed into mixed oxide fuels*" challenge and we have the technical know-how to do that safely indeed. For example, used, hot, fuelrods are mostly stored "*on site*" in tanks of water when they are removed (*robotically*) from the reactor. Water safely absorbs the rapidly-emitted neutrons that are most injurious to human health. In so doing, the water becomes "*neutron-heavy water*" which, in turn, has several useful industrial applications.

Most reactors store used-fuelrods in water on site for three years or more and then they are placed in sealed canisters and stored in specially designed repositories until they are reprocessed for heavy metal extraction and mixed-oxide fuel production. The Canadians have stored 45 years of cumulative nuclear electricity waste in very small containers on various plant sites; there it will stay until it is reprocessed into mixed oxide fuels. Even better news, the Japanese have solved the need for spent fuel storage. They have developed a small reactor (*which incorporates hydrogen production and desalination*) that has appropriately dispersed and configured the *very small quantities of fuel* it uses so that it does not need re-fuelling for at least thirty years, by which time all of its fuel's potential for producing heat has been entirely used.

The latest collaboration between Microsoft and various Japanese Government and industrial agencies is looking at developing a 100-year operational life for a new configuration of the fuelrods. These will also be designed to result in negligible radioactive waste (*Nuclear Power Daily, 2010*).

In a light-water reactor's fuelrods operational life, some lighter fission products are produced that poison and stop the neutron chain-reaction. These are called "*actinides*"; two of which are quite nasty: they emit radiation rapidly and dangerously, for the thirty to sixty year period most harmful to humans and other biological life. But these substances present a miniscule quantity of material, less than three percent of the material in a spent fuelrod, and they can be chemically separated from the remaining, 97 percent recyclable, metals. Responsible management now ensures that such rapidly-emitted radiation sources are not allowed to interfere with the living biosphere. Such radiation sources are isolated when they cool and are now stored in underground storage facilities, encased in glass, cement or clay or synthetic rock (*which emits negligible radiation),* and these forms of disposal effectively isolate them from causing harm while they decay to background radiation levels (*now thought to be a period of ninety years).*

Strictly monitored, international, management conventions have been established in relation to disposing irradiated, and radiation emitting, reactor materials safely. Even so, some yet-to-be adequately-explained practices remain and it is right for the public to remain vigilant about this particular issue. For example: there are reported to be steel drums of waste left in the open air on Siberian plains and on a slab of cement beside a Prairie Island flood plain in the United States (*source: Nuclear Power Daily*). Both sites are "*off-limit*" to the public but further enquiry is required and better management may be immediately required if, indeed, the material is hazardous. It is possible the radiation symbol is being used on the drums to deter vagrants.

Responsible authorities categorize waste into three levels of intensity: **Low, Intermediate and High Level Waste.** (*A very good source of information is the compilation, Australia's Uranium, 2006. Some of the facts of the matter are startling: the most dangerous radiation humans now receive is that associated with magnetic resonance imaging, dental and medical x-rays and radiation supported chemo-therapy. Yet, all of these treatments are now regarded as de rigueur and safe, by the public at large. Nuclear electricity production has a much better safety record than nuclear medicine.*)

The very little percentage of genuine waste (***three percent*** *of the material remaining in a conventional used fuelrod: light-weight fission products and short-lived, heavier than uranium, transuranics*) that is currently uneconomical to use or is particularly dangerous, either as fuel or for medical isotope production, can be either disposed of permanently through high temperature incineration and "*synroc*" encasement, or disposed through "*synroc*" encasement and geological burial, alone. When radioactive material is incorporated into "*synroc*", it results in less surface-radiation emissions than that which is produced from the original uranium ore or from other radiation-emitting rocks like granite (*Hardy, pers. comm., 2008*). But any radioactive material, even without encasement, if buried deep in a stable geological repository and back-filled with clay or salt, will no longer have the potential to contaminate the biosphere. The phenomena of *"transmutation"* (*the process of atomic and elemental change that happens as radioactive energy is transferred from atom to atom, diminishes in intensity with distance*) shows that buried radioactive material's energy can be fully absorbed by a meter of dense clay or salt. The only danger to the biosphere that might happen from that form of storage is when waterborne *organic* particles pass through the soil/clay or salt. They are at risk of becoming irradiated and so, might then be transferred to living things through water ingestion. It is better to minimize and avoid this hazard: that is why sub-surface storages for contaminated materials are deep pits, located below water tables (*described technically as stable, deep geological repositories*), or dry, above-ground, cement, shelters. But to date, (*Australia's Uranium, 2006, p121, records that*) *"no **permanent** High Level Waste repository has been constructed anywhere"* in the world because there is so little of it in quantity and interim storage has seen its radioactivity decay to manageable, background levels (*enabling recycling*) within decades. Careful enquiry suggests that radioactive waste should be one of the least-to-be-feared forms of modern industrial waste, but it would appear that only those who have seriously studied nuclear power science know this with certainty and a ridiculous situation has emerged where the opinions of such persons are not trusted because they are *"associated with the industry"*.

Low Level Waste is material that may have been contaminated with some radioactive energy. It consists of the laboratory equipment such as gloves and gowns, rubbish bins, pens and paper, and the wipes and tongs that have been used by people who work in nuclear science industries. It is material that will show no signs of more than background radiation after ten years. It is material that does not, inherently, emit radiation. This, generally innocuous, material is, by volume, more than eighty percent of all of the nuclear industry's waste. Such material can be stored above ground for a period of up to ten years, by which time its emission status will be equal to or less than most natural background radiation. It can then be buried for compost or burnt safely. Low Level Wastes do not present a significant management problem.

Intermediate Waste is material that contains some levels of gamma and beta emitting contaminated material, so it usually requires remote handling and longer storage, up to between thirty years and three hundred years before its emissions degrade to background radiation levels. It does not require "*synroc*" encasement. Intermediate Level Waste includes materials such as dismantled reactor vessel parts and the cement used to house a reactor vessel. This form of waste is safely managed today, on site or in specially designed storage facilities, by shielding and containment in above-ground, waterproof structures made of cement. Intermediate waste amounts to between four to five percent of all the residual radioactivity produced in a nuclear power plant. It represents only seven percent *by volume* of all a reactor facility's waste.

The Commonwealth of Australia has produced a wonderful guide to the safe storage of both low-level and intermediate-level waste materials called *The National Radioactive Waste Repository, Draft Environmental Impact Statement Summary.* Reading this reveals how clever the Muckaty Station traditional owners were to accept a generous offer of significant finance to build a Low to Intermediate Level Waste storage facility on their land. Although the station is renowned for its people's production of innovative art, it is always challenging to make desert country economically productive, and the proposed waste repository, carefully managed according to the guidelines developed here in Australia, will do just that!

The genuine **High Level Waste** is a very small amount, about three percent of the spent fuel rod. It is called High Level Waste because it is able to continue generating thermal power through radiation emission, i.e. it remains *"very hot"* for several years, and it is not able to be used easily for further power production. The rest of the fuel is now stored in preparation for recycling and reuse in electricity generation (*the United States announced this year that it would commence mixed oxide fuel production and Russian, British, French, Swiss, Japanese and Germans already pursue this recycling option*). Spent fuel-rods contain about 95 percent plus of re-useable, stable, depleted uranium and two percent of plutonium and uranium-235. The residual three percent (*radioactive actinides*) could also be *"burnt"* to produce electrical power, but the technological costs associated with building new, high-temperature-resistant, infrastructure have been prohibitive. It is quite safe, and thought cheaper, to dispose of the actinides in glass or synroc encasement and then bury this in geological repositories. If this small part of the High Level Waste were utilized to produce power, the high temperature burning would cause its emission rates to be accelerated and its hazardous emission potential could be significantly reduced. The end-product, the residual *"burnt waste"* would also be suitable for either *"synroc"* encasement and burial, or even above-ground storage, until it had decayed to the equivalent of background radiation levels.

For those of us who have feared nuclear waste, it is worth noting that the volume of High Level Waste is very little: 3.5 *million times less* than the waste produced by burning coal to generate an equivalent amount of electricity (*Beckman in Keay, 2003*). Globally, only 12,000 tonnes is produced each year compared to 25,000,000,000 tonnes (*25 billion tonnes*) of carbon waste from the use of fossil fuels (*Australian's Uranium*, 2006, p21) and the bulk of this HLW is recyclable. Storing and disposing of genuine High Level Waste material, the nuclear science community agree, has not been, nor will it ever be, a prohibitive or physically large, undertaking.

The eighth lie

A great many people believe that some isotopes of plutonium and uranium will emit radiation for hundreds of thousands of years and so will be forever dangerous. They actually believe this of all radiation-emitting sources. This is heavy propaganda. A spurious association has been deliberately contrived. Create a culprit and advise it is long-lived, forever at work, and people will reject it instantly and with great conviction! **The truth of the matter is that long-lived radioactive isotopes emit radiation *so slowly* that you can hold them in your hand with impunity.** Copper, gold and silver all have such isotopes and we wear jewelry crafted from these metals every day of our lives!

To properly understand the essence of this misinformation we need to understand what the *"half-life"* concept means. A half-life is used to measure the time it takes for radioactive material to decay to stable material. In one half-life, half of the material will decay. In two half-lives, 75 percent of the material will have decayed. In three half-lives, nearly 90 percent of the material will have decayed, and so on. Uranium-238 decays very slowly. It has a half-life of 4.47 billion years and eventually decays to form lead.

It is the short-lived radioactive isotopes, those that emit radiation rapidly, that are dangerous to the living biosphere. Those that are particularly dangerous to humans are the isotopes with a half-life of less than thirty years. If ingested or if humans come into continual, close, physical, proximity to these substances, it is possible that the body's *hormesis* response and *aposteosis* mechanisms could be overwhelmed and some form of radiation poisoning could occur. For this reason, all nuclear facilities are managed to prevent any form of continual, close proximity to such materials from occurring. These substances are contained and isolated from the environment until their radiation-emitting capacity is exhausted: this can happen if they are immersed in water for a few years; or, if they are remotely reprocessed, incinerated at high temperatures, and/or encased in glass or synroc, then buried. Any and all of these disposal options minimize the hazard. Once they are deeply buried in geologically stable repositories, they are no longer hazardous to the living biosphere.

The synroc (*synthetic rock*) containment process, an invention by Australian Professor Ted Ringwood, is an ideal treatment and a wonderful tribute to his life's work.

Synthetic rock, "*synroc*", encasement is a mineralization disposal method which binds radioactive material into a crystalline structure so that it does not leach into the surrounding environment. The non-leaching feature is important because living creatures could be exposed to a radiation hazard if organic particulates suspended in water were to "*leak*" from the containment. Synroc is impervious to water and this form of contamination, water transferring the radioactive isotopes that may have adhered to organic particles, is prevented. It is a wonderful development. You can view it in Australia at the Lucas Heights public display room. It looks a little like graphite, but has a slightly-more rubbery texture.

Other spurious rhetorical positions

As the preceding discussions show, it would be wise for nations genuinely concerned with providing their populations with enabling wealth and energy security to ***immediately*** commence employing small and large, efficient, nuclear power generators. There is no need for them to develop fuel production processes: the nuclear-advanced, developed world is morally obliged to share this technology under affordable arrangements so that all of the nations of the world can realistically combat *"climate change"*. Accepting this position will negate the fears many have held about weapons being developed by disaffected member of the global community. It would be wise for as all to plan how best to apply the plentiful, inexpensive, power these plants will produce to create opportunities for useful industrial applications and local employment. It would be wise to commence planning for food production and reforestation by harnessing nuclear power's portable and mobile potential for supplying fresh water by purifying oceanic or brackish artesian sources. We could also, as the Japanese have commenced to do, commence to manufacture clean hydrogen gas for the creation of a much-needed, non-polluting, generation of hydrogen gas-powered heavy-duty haulage for rail, sea and air transport.

But, there is entrenched resistance to such courses of action in many areas. Those who benefit financially from fossil fuel often exhibit all the predictable behaviours of the opponents in a *"classical scientific revolution"* (*Thomas Kuhn, 1968*). They will fight to the death to retain their power and economic influence. There is also resistance by passionate, but poorly informed, environmentalists. The other notable resistors are *"puppet politicians"* who are characterized by either having ignored the advice of nuclear scientists, or else they have been deceived by the myths and money produced by those who have been interested in obscuring the true value and opportunities uranium and nuclear power will provide to the *entire* world. The rhetorical positioning, used to rationalize such persons' failure to support the nuclear power alternative, must be carefully examined to expose and refute the false reasoning that dominates public discourse on nuclear power matters.

One persistent objection is made in relation to the process of *enriching* uranium oxide. Very highly enriched uranium can be used for bomb production and this is the reason why enrichment processing is, by international agreement, restricted to only a very few countries; all Government owned and supervised by military security.

While it is true that enriched uranium ore is not an essential requirement for generating electrical power (*the Canadian CANDU reactors were designed to use un-enriched uranium oxide*), the scientists with whom I have communicated believe a more efficient fuel is obtained from uranium-235 when it is enriched to a concentration of three to five percent of uranium oxide. This is because one of the assured by products of enriched uranium fuelrods is extractable plutonium. Plutonium, in turn, is needed for enabling all of the depleted uranium ore body (all stocks of uranium-238) to be used as mixed-oxide fuels in the future.

Enriching uranium in Australia, now one of the world's largest suppliers of uranium oxide, will enable Australia to manufacture fuelrods suited to the mini-power generators and desalinators that will particularly help isolated, regional, dry Australia, and most of the Southern Hemisphere, achieve greater prosperity. The question is, can further enrichment happen safely and not result in more weapons manufacturing? Yes. If access to uranium, both the rare earth resource and processed and used fuels, remain strictly guarded and "*off limits*" to all those who are not entrusted (*by the international community*) with its production and distribution.

Enrichment plant (*fuel production*) operations around the world are governed by the *Nuclear Non Proliferation Treaty (NPT)*. The signatories, 190 states in total, all, either undertake not to produce nuclear weapons, or the states that have them undertake not to use them. India, Israel, Pakistan and North Korea have not signed this treaty, we can only hope that sanctions that prevent their being able to share their technology through global markets will create a need for them to become signatories. But, it is worth bearing in mind, the proposed sale of Australian uranium oxide to India for civilian power generation, via processing in the United States, has been subject to even more advanced scrutiny than that provided by the older NPT (its governing clauses are articulated in *The India Specific Safeguards Agreement)*. This agreement was developed under the auspices of the International Atomic Energy Agency. Instead of the sale of uranium oxide to India being thought a dreadful, threatening, development, it is one that will allow India to produce nearly three quarters of its electricity needs from non-polluting sources before 2050. With access to reactor fuel, the Indian negotiator, Mr Kakodakar, says they will more than double their electricity generating capacity, from 4,120 megawatts to 10,000 megawatts, by 2012 and, in so doing, significantly reduce their dependence on imported oil which currently supplies 70 percent of their energy use (*Straits Times, 2009*). This is a marvelous, empowering development for India (*and for the health of our global atmosphere*).

The real impetus behind the concern to limit enrichment opportunities is that the nations who have been *"hell-bent"* on acquiring nuclear power capabilities are seen by many to be the least politically-stable. Yet, these same nations can also be viewed as those that were most threatened by the atomic weapons capacity acquired most of Europe, the former Soviet Union, Israel and the United States. A huge challenge is, how to defray this fear?

The policy positioning by President Barak Obama, calling for the global abolition of *all* nuclear weaponry, might remove some of this international anxiety, but it is hard to imagine the world superpowers actually achieving this entirely in the near future. As I write (*April, 2010*), the President of the United States has endorsed the recall of thousands of atomic weapons in a new initiative called *"mega-tons to mega-watts"*, and he has pledged that the US will never use nuclear weaponry against non-nuclear states. A great deal of trust is now required for the world to maintain this peaceful impetus and for non-nuclear weapons nations to abandon their nuclear-weaponry aspirations: this trust can only be created by generous, benevolent, actions, respectful of more equitable global prosperity and technology-sharing, not *rhetoric* alone.

In the meantime, an immediate opportunity has surfaced for the nations that have **no interest in developing nuclear weaponry** (*Australia, New Zealand and Canada are some*) to be allowed to develop enrichment capability and supply leased fuel to developing neighbours - it is easy to contaminate fuel so that that weapons capacity is nullified. There would be no hypocrisy in this position for Australia does not possess, nor does it want to possess, independent nuclear weaponry. We have, through our alliances, second-strike capability and that, until the entire world disarms, is deterrent enough.

In relation to other nations sneakily attempting to enrich uranium, Australian scientists have invented several clandestine methods for monitoring enriched uranium production and these methods are now used mines and in all internationally-co-opted utility developments (*mentioned in* Hardy's, 1996, reprinted 2008, book, *Enriching Experiences*). These methods detect where, and when, super-concentrated enrichment might be happening.

The inherent safeguard designed to preclude diversion of uranium to weapons manufacturing, relies on what is called the *"timeliness criteria"* (*see Nuclear Safeguards and the International Atomic Energy Agency's web site*). With frequent and varied inspection regimes, predicted isotopic outputs can be compared to the actual outputs over time, and so, any diversion of material for greater enrichment can be easily detected. This was how Iran's rule-breaking diversion program was detected. But it is good to see the tension behind the Iranians' wanting access to nuclear power and enrichment technology gradually being defrayed: i.e. the Iranian President's announcement that his country would happily accept fuelrods from internationally accredited nations and use these for electricity production, not weapons manufacturing. Iran, like all of the world's nations, is entitled to have access to nuclear powered electricity. It is to their credit that they see benefit in developing this capacity independently of dependence on other nations. One can only hope and trust that the vulnerability they feel by living close to ancient, tribal, and modern enemies who possess nuclear weaponry, will abate when all nations embrace nuclear disarmament as a most humane and civilized extension of their national and human identity. Until then, it seems best that the nation states that are technologically-able to produce fuelrods continue to contaminate enriched fuel to prevent any potential purchaser/leaser from recommitting the fuel to atomic bomb production. The IAEA, and the Australian program under John Carlson, is doing outstanding work in this area.

If politically stable and technologically-advanced nations, like Australia, do not assume responsibility for the nuclear fuel cycle and share nuclear power generation opportunities in their regions, a great deal of their potential for influencing outcomes, such as peaceable regional relationships and prosperity development, will be forfeited. Directing energy to creating prosperity is a hugely stabilizing ingredient, perhaps the single, biggest, disincentive to aggression humanity possesses. So, it is, in fact, in the pursuit of world peace and our global security that stable, prosperous and generous nations, like Australia, develop this engineering and technical capacity, as soon as possible.

Unshackling the entire world from fossil-fuel dependency and allowing humanity to benefit from nuclear power science, meets with other forms of rhetorically-plausible resistance, but our not changing-over and continuing to burn fossil fuels, is far more reprehensible. To continue burning coal to generate electricity when a clean alternative like nuclear power is available is ecologically irresponsible. It is also squandering the precious coal and natural gas/petroleum resource, which, if it is not burnt, is able to provide Australia (*and the entire world*) with the complex carbon molecule feedstock needed to manufacture plastics, synthetics and polymers for many generations to come. Hydrogen gas is the best contender for supplying non-polluting fuel for heavy-duty air, sea and rail transport in the future. Hydrogen, being the smallest element, will "*sneak*" through almost all known containment materials, except man-made polymers. A thin polymer film in a steel canister enables us to contain hydrogen gas. If humans continue to burn coal we deprive future generations of options, such as safe hydrogen containment (*polymer production*), and all manner of other, useful synthetic products.

Furthermore, the cost of burning coal must factor in: a) the exorbitant and steadily growing cost of transporting millions of tonnes of coal to power stations. These costs escalate with every miniscule rise in the cost of oil. And b), the cumulative, deleterious health cost born by humanity coping with the toxic pollutants associated with burning fossil fuels.

Similarly, burning natural gas to produce electricity is just plain wrong. Natural gas is an incredibly finite resource. It has no replacement for its use in industrial furnaces or welding and the entire, industrial, developed, world depends upon these processes.

So, it is critical to judiciously manage the remaining fossil fuel resources and try to prolong the time we can have access to their indispensable-to-industry applications. In relation to atmospheric health, clean nuclear power is eminently preferable to burning coal or natural gas for electricity production or transport uses. It will provide the high-grade, base-load, electrical energy that will allow us to cleanly manufacture hydrogen gas (*by running an electric current through water*) and recharge our own electric cars, inexpensively, while they are parked in our garages.

With regard to those who object to nuclear power on economic of grounds: since generation two and three reactors have been re-licensed to operate for another twenty years in the United States *(giving these utilities a sixty- to eighty-year-plus, not a forty-year-plus, life span: other utility-life extensions are now happening in Taiwan, Belgium, Germany and elsewhere)*, all of the past hostile predictions and comparisons made about nuclear power costs are redundant **and wrong**. The bulk of these extra costs were incurred during the primary research phase and in building infrastructure and these costs are now able to be dispersed over a far-greater utility lifespan. Uranium oxide fission power is rapidly becoming the least-expensive of all current electricity producing alternatives. And when mixed-oxide fuels commence to be used, they will cost even less as these fuels are derived almost entirely from recycled fuel metals.

Concluding thoughts

One is forced to conclude, after this examination of the ideas used to justify opposition to nuclear power development, that although the positions and statements examined above are rhetorically plausible, many are, in fact, nonsense. That they have been believed for more than fifty years by many millions of people has huge political implications. Without an adequate knowledge base, one is incapable of determining sound advice from that which is spurious. Our leaders, in future, will have to be very well-informed and well-educated indeed. **Energy security is perhaps the most critical question any modern nation faces and for any elected official in the developed and developing world to remain willfully ignorant of the very basics of nuclear science, now amounts to culpable political negligence and a betrayal of trust.** Advocating nuclear power production and its allied *"fuel cycle"* industries is not mere opinion. This is a logical, fact-supported, position, worth a great deal more credence than mere opinion. As Bernard Cohen has said, *"the pattern is clear, the more one knows about nuclear power, the more supportive one becomes of it"* (Cohen, 1994, p44).

References

AFP, 19th August 2009. *Sharp rise in India nuke power,* Straits Times, Singapore Press Holdings Ltd, www.straitstimes.com. Forwarded by Ian Ng, Fiji.

ANA, 2005. *The Facts on Nuclear Science, Uranium and Nuclear Power,* Sydney, Australian Nuclear Association, PO Box 445, Sutherland, NSW.

Atkinson BMC, 2008. *An Emergent Communication Strategy to Accelerate the Transition to Ecological Sustainability,* doctoral thesis, School for Environmental Research, Institute of Advanced Studies, Charles Darwin University, Northern Territory.

Australia's Energy Options, October, 2005. Future Directions International Study.

Australia's Uranium: Greenhouse Friendly Fuel for an Energy Hungry World, November, 2006. The Parliament of the Commonwealth of Australia, House of Representatives Standing Committee on Industry and Resources.

Cohen BL, 1990. *The Nuclear Energy Option: An Alternative for the 1990s,* Sage Publications, Inc, 233 Spring Street, New York, London.

Comby B, 1994. *Environmentalists for Nuclear Energy*, TNR Editions.

Cote A, 2005. *A Class of Lightweight, Rigid Polymers that could be Useful for Storing Hydrogen*, University of Michigan, Nanotechwire http://nanotechwire.com/news.asp?nid=2604

Davis D, (Devra) 2007. *The Secret History of the War on Cancer*, Basic Books, a member of Perseus Books, Park Avenue, New York.

Energy Resources Australia, *ERA History*, 2006, http://www.energyres.com.au

Hardy CJ, 2006. *A Cradle to Grave Concept for Australia's Uranium*, Sydney, Four Societies Meeting, Eagle House, Milson's Point, 22 February 2006.

Hardy CJ, 1996 & 2nd edition, 2008. *Enriching Experiences: Uranium Enrichment in Australia 1963 – 2008*, Glen Haven Publishing, PO Box 85, Peakhurst, NSW, 2210.

Hore-Lacy I, 2005. *Nuclear Power: Current World Status and Future Trends*, Australian Nuclear Association, Sixth Conference on Nuclear Science and Engineering in Australia, Sydney, Weston Hotel, Australian Nuclear Association.

Hunt FK, 1850. *The Fourth Estate*, London, David Bogue.

Keay C, 2002. *Nuclear Energy Gigawatts: Supporting Alternative Energies*, PO Box 166, Waratah, NSW 2298, The Enlightenment Press.

Keay C, 2003. *Nuclear Common Sense*, PO Box 166, Waratah, NSW 2298, The Enlightenment Press.

Keay C, 2004. *Nuclear Radiation Exposed: A Guide to Better Understanding*, PO Box 166, Waratah, NSW 2298, The Enlightenment Press.

Keay C, 2005. *Nuclear Energy Fallacies: Here are the Facts that Refute Them*, PO Box 166, Waratah, NSW, 2298, The Enlightenment Press.

National Archives of Australia, 1969. *Submission 759, by the Minister for National Development on the Establishment of a Wholly Commonwealth-owned Nuclear Power Station, 1969* Cabinet Documents: 69.

National Radioactive Waste Repository Draft Environmental Impact Study, PPK & Commonwealth Department of Education, Science and Training, 5 Star Press, Adelaide, South Australia.

Nuclear Australia: Newsletter of the Australian Nuclear Association Inc, 2005 to 2008 inclusive. PO Box 445, Sutherland, NSW.

Nuclear Power Daily, an online newspaper providing news from around the world pertaining to civil nuclear power. It is an excellent resource, www.nuclearpowerdaily.com/reports.

Nuclear Safeguards and the International Atomic Energy Agency, accessed 12/09/08 http://books.google.com.au/books?id=dmv-LfLNzrEC&pg=PA70&lpg=PA70&dq=enrichment+safeguards& source=web&ots=G_X6ZDtzM&sig=WKBE6nBVd8aYDBWnd1 VBoLKqKUg&hl=en&sa=X&oi=book_result&resnum=5&ct=res ult#PPA59,M1

Nuklearforum, August, 2009. www.nuklearforum.ch/nuclearplanet.

Radiation and Life. Uranium Information Centre Ltd. GPO Box 1649N, Melbourne, Victoria, 3001.

Straits Times, 19th August 2009. *Sharp rise in India nuke power, AFP*, courtesy of Ian Ng.

The Legend of King Solomon's Mines; The Sleeping Beauty and Carbon Dating; The Legend of Jolly Rogers; Australian Nuclear Science and Technology Organization, http://www.ansto.gov.au

WANO, 2007. *What is WANO?* World Association of Nuclear Operators www.wano.org.uk/WANO

Wikipedia, www.en.wikipedia.org/wiki/, www.npi.gov.au/database/substances-info; www.scorecard.org.chemical-profiles/summary; www.nap.edu/openbok.php?recordid; www.info/anu.edu.au/hr/OHS/Hazard Alerts/ for searches relating to Cadmium_poisoning; Arsenic; Arsenic_poisoning; Selenium; Tellurium; Boron_trifluoride; and hydrofluoric_acid.

Williams DR, 1998. *What is Safe? The Risks of Living in a Nuclear Age*, Department of Chemistry, University of Wales, Cardiff, UK.

Letter two
Ecological sustainability and nuclear power

The quintuple bottom line and
the ecological sense of nuclear power matters

Dear Friends,

There is a qualitative dimension to advocating a nuclear powered energy infrastructure for our entire world, and that is, that it is the *"right thing to do"*.   Nuclear Power will, as a corollary to massive reforestation, enable us to effectively combat climate change and become genuinely ecologically-sustainable.   We can have confidence that this determination is correct by referring to an extension of the *"triple bottom line"* principles to a *"quintuplet"* of qualitative criteria to guide our development policies and infrastructure-development proposals.

To help you to be certain that the determination to advocate nuclear power production is, in almost all instances, simply the best, sustainable, high-quality source of electrical energy, I need to share more fully the results of my own research and thinking on this issue.

Firstly, we need to improve our definition of what constitutes ecological sustainable development.

Defining ecological sustainability

There are many definitions of what constitutes *"sustainability"*, *"ecological sustainability"* and *"ecologically-sustainable development"*, however, the terms can all be garnered towards a similar destination.

For example, *"ecologically-sustainable development"* implies the requirement of judicious management - dominion with wise stewardship - to achieve and maintain the goal of *"ecological sustainability"*. I describe *"ecological sustainability"* as the term that represents **nature's ability to self-perpetuate life** through supporting increasingly diverse interrelationships between ever-diversifying, fluctuating communities of co-dependent species. These complex interdependent communities, with their legacy of modifying the physical parameters of their local environments, help to create the ecosystem functions that collectively create the conditions we can describe as *"a biosphere amenable to the evolution of life"*.

However, this definition still falls short of a satisfying and accurate description of what would constitute *"ecologically-sustainable development"*.

Management, dominion and stewardship are important dimensions of the comprehensive definition *"ecologically-sustainable development"* for the inclusion of those words recognizes the potential humans have in thwarting or assisting *"ecological sustainability"* to manifest. Recognizing these extra dimensions of responsibility, Kerkhoff and Lebel (2006) nominate a *"classic"* definition of sustainable development as that being made by the *World Commission on Environment and Development (United Nations, 1987)*. The Commission described it as development which *"meets the needs of the present without compromising the ability of future generations to meet their own needs"*.

Kerkhoff and Lebel's own personal *"approach"* maintains *"sustainable development"* as *"the process of ensuring all people can achieve their aspirations while maintaining critical ecological and biophysical conditions that are essential to our collective survival"* (Kerkhoff and Lebel, 2006).

This latter definition, which emphasizes *"maintenance"*, still falls short of reflecting the reality of the situation we face globally. We desperately need to **restore and replenish the biosphere's life-supporting propensity**. Desertification is *"in train"* globally (*Atkinson, 1990, 1992*). This macro-climatic process is epitomized on the micro-climatic scale by the *"edge-effect"* that happens when we encroach upon a forest. Where shade and moisture are no longer afforded to protect vulnerable species from direct sunlight, harsh winds and the greater temperature threshold variations at the edges of forest subject to severe disturbance (*such as, logging, fire and wholesale clear-felling*): these encroachments combine to cause *"dieback"* on the perimeter of the disturbance (*Florence, 1979*). This modest *"edge-effect"* reduction in diversity has ramifications because the dying species' dependent species, when deprived of their hosts, food or simple, protective, habitats are, in turn, subject to stresses that can result in death. Habitat disturbance and deprivation creates a chain-reaction of destruction can be described as the process of *"ecosystem simplification"* (*Atkinson, 1992, www.specialistwritingservices.com.au*). This unrelenting process is happening in innumerable, diverse ecosystems, globally.

Furthermore, critical to any progress towards achieving the goal of socio-cultural *sustainable* development is the recognition that, globally, our combined exploitive activities are not now ecologically-sustainable (Boyden 1990; Boyden 1992) and any suggestion of being able to create such a world must necessarily include both **minimizing the damage we do to the life-sustaining propensity of the biosphere and, immediately, seek to repair and replenish this propensity through bio-diverse strategic plantings aimed at restoring complex, life-sustaining biomass at every possible opportunity**. We cannot afford the time now needed for the ages-old processes nature has (*like wind-blown re-seeding*) to repair the monumental damage of our post-industrial activity: human intervention and strategic management are desperately needed to accelerate this process.

Thus, my definition of *"ecological sustainable development"* embraces the need to not only **manage the natural world's resources wisely and facilitate its natural processes; it also includes the imperative to preserve, restore, replenish and even, enhance, bio-diverse biomass re-establishment**. I particularly emphasize *"bio-diverse biomass"*, because restoring global biomass will, undoubtedly, help us arrest and even begin to reverse *"climate change" (see definition entire, page 98)*.

Arresting climate change

Arresting and reversing climate change is a critical component of achieving genuine ecological sustainability. Climate change is not only caused by burning fossil fuels to provide transport and generate electricity: it is caused by using energy inefficiently and producing too much waste heat. To eliminate this contributor to climate change, our energy-use patterns *must be* circumscribed to compliment the natural world's capacity to negate ill effects of the extra-somatic energy burden human activity has imposed on the biosphere. Hence, learned ecologists like Professor Stephen Boyden (1990, 1992) and others, suggest that we must reduce human energy consumption, that used for electricity production, industrial end uses and transport, globally, by as much as seventy-five percent. This challenge presents industrial designers with phenomenal opportunities to innovate new, sustainable, tools. This opportunity is one of the main reasons we can be confident that genuine attempts to create an ecologically-sustainable world will enable us to prosper economically; and this is certainly not the only opportunity.

Climate change is also caused, or significantly exacerbated, by removing forests.

In the late 1980s, while studying for a Master of Philosophy at the University of Wales, I monitored environmental news stories from all parts of the globe for 18 months and I noted a significant anomaly. At that time the *"greenhouse theory"* was *"in vogue"* and it proposed that we would experience rising temperatures all around the world. But that was not what was happening. Not only were hottest temperatures on record being reported from all around the world, less frequently, but consistently, the coldest temperatures on record were also being recorded. I reported this anomaly, that we were experiencing a breakdown in the stable parameters of climatic temperature thresholds, hotter *"hots"* and colder *"colds"*, at an international environment conference in Oxford, 1990, and later at a Conservation Biology conference in Brisbane, 1992.

While the hotter temperatures were expected and explicable through the *Greenhouse Theory* idea (*because burning fossil fuels adds a blanket of carbon, gas and particles, to the atmosphere, causing heat to be retained and preventing it from escaping into outer space, consequently warming the planet. We see this opaque blanket all over the Northern Hemisphere because there, we can see "air": the legacy of continuous fossil fuel and forest/stubble burning*) that the *"coldest temperatures on record"* were also happening, was not explained by this theory. I tentatively suggested that this anomaly (*hottest **and** coldest temperatures*) was able to be explained by our plundering the world's forests, creating deserts, all around the world. Where we had not totally cleared the forests, we had significantly *"simplified"* their species composition. We could call this phenomenon, the *"simplification"* of ecosystem, results in desertification. It is able to explain the anomalous coldest temperatures on record surfacing because deserts are characterized by very hot temperatures during the day and freezing temperatures at night. I am certain that regardless of the impact of increased fossil fuel burning, **climate change is explicable through the practice of removing complex forest, alone**.

The advent of rapid, widespread, desertification or *"ecosystem simplification"* explains the extension of temperature thresholds now observed, globally. How?

A single tree can influence micro-climates. Its canopy of leaves create a moist, cool, shady umbrella of protection from the heat of the day, breaking the energy impact of falling rain and hail and reducing wind attrition of soil; clumps of trees even lifting the desiccating wind, causing it to blow above the tree tops and not close to the ground. A single tree will also channel rain water deep into the soil through its roots, replenishing aquifers and underground water tables, contributing to greater moisture retention during hot summer months. A tree's leaf litter decays to form humus which also helps the soil to retain moisture. Thus, a single tree can help prevent local erosion and creates the conditions (*food, moist soil and shady, cool and moist air under the canopy*) that will support other life forms flourishing. It also contributes to a healthy atmosphere by producing fresh oxygen and water vapor as a by-product of absorbing and converting the sun's energy and atmospheric carbon dioxide into carbon-atom-rich timber through its leaves in the processes of photosynthesis and transpiration. If a single tree can alter its microclimate in this fashion, then undoubtedly, a whole forest will alter macroclimates. A forest can absorb a great deal more energy and water and convert a great deal more carbon into woody fibre. In so doing, a forest will create a significantly cooler tract of land or, in cooler climates, its protective canopy and energy absorption propensity will create a more temperate, warmer environment. Hence, forests *"ameliorate"* climatic extremes.

All around the world, the cooler tracts of land and moisture-laden air above forests cause the adiabatic differentials that result in rising and falling pockets of air and wind movement from hot, low air-pressure zones towards the denser, cooler, high-pressure air pockets. Where warm and cold air zones intermingle, rain *"on the front"* is one of the likely consequences.

Where hot, moisture-laden air rises, condensation will also happen, and at predictable times of the day, often in the late afternoon. In this fashion, collective forest ecosystem functions contribute to stable, predictable, rainfall regimes. These patterns are reliable and can be predicted because they respond to the intensity of the sun's energy, which, in the tropics, the birthplace of the clouds, maximizes between midday and three pm each afternoon. It is a very old adage indeed that reflects this adduced knowledge: the simple, ancient saying, *"trees bring the rain"*.

Removing a local forest significantly interferes with the predictable rainfall patterns of that locale. It also removes all of the other advantages forest cover provides, like: stabilizing water tables; the vital contribution a forest's leaf litter makes to soil-formation and soil-protection; the effects of shade and shelter on temperature, humidity and air-flow regimes; and their other, complex, life-supporting, habitat-providing roles.

A simple extension of logic enables us to realize that removing *substantial* forest, regionally, causes *substantial regional* climate change. Hot or cold, dry winds replace moisture-laden air flows, shriveling life. And when rainfall arrives it simply results in floods and huge loss of top soil, and so, the region's potential to support abundant life through soils - the result of many thousands of geological episodes - is significantly diminished.

The extension of temperature thresholds that happens with large-scale clearing of forest, and its resultant desertification, is particularly threatening to whatever *remnant* vegetation remains and accelerates simplification's chain-reaction of death.

For example, many plants, and plant communities (*complimentary species associations – those which are happy to live together*), will die if they are subject to an unseasonal frost or an uncommon hot spell: their cellular metabolism fails when ice forms or desiccation happens and single plants can be severely damaged and even die. If this happens often enough or at an untimely moment, before seed is set, whole communities will perish: some permanently, because seeds can be viable for only a few moments, days, or weeks.

With no seed, those plants **will not grow** in that location again.

Removing forest globally, on the scale we have, causes incredible harm and the ensuing *"simplification"* of species diversity represents *"a reversal of the process of evolution"*: evolution being the amazing process which provided for greater complexity and species diversity manifesting to compliment and counter-balance the changing conditions presented by an increasingly, ever-more, complex, biosphere.

Thus, climate change is not only the result of burning fossil fuels for electricity production and our industry and transport needs, it is also a direct result of clearing the world's forests: we have lost 50 percent outright and severely degraded another 30 percent (*World Resources Index*, 2000). (*There is another significant cause of climate change: ozone depletion, but I discuss this towards the end of this letter.*) Hence, one of the genuine solutions to climate change, besides ceasing to burn fossil fuels, reducing the amount of extra-somatic we manufacture and ceasing to pollute our oceans and atmosphere, is the **replenishment of bio-diverse biomass** all around the world.

In other words, we humans, of every nationality, need to quickly replant the world's forests, emulating the patterns and species associations pristine natural conditions provide, and conserve as much as remains possible. Our planting efforts could, for example, imitate seed fall and natural dispersal patterns. Seeds can be distributed by insect and animal activity and by the wind. Copying nature includes creating diverse plantings of single species clumps intermingled with other, diverse, plant communities. This is preferable to single mono-cultural plantations of local or introduced species because diverse communities are more robust. They have greater resistance to disease and create a greater tolerance of other, ecological permutations. Another plan could be to use pioneering species, like nitrogen-fixing Acacias, to create shade, underneath which more vulnerable species can be established. With the addition of fertilizers made from effluent, rainforest can be re-established. In all plantings, *habitat creation* opportunities to provide wildlife with food and shelter should be maximised, with special attention paid to providing wild fauna with access to water.

To replant forests, we need access to healthy seeds. At this very early stage of the global, human-driven, socio-cultural quest to achieve ecologically-sustainable development, it is essential to protect the world's remaining old-growth forest for its critical bio-diversity repository value. For example, a single, mature tree in an old-growth forest can sustain and harbour more than fifty species of life. These species can all be permanently lost when a tree is felled and *"we have simplified that ecosystem"*. To appropriately restore ecosystems, we need to provide for species migration and/or be prepared to transplant entire communities of plants and animals and micro-organisms. Removing the tiny remnant of genuine old-growth forest that is left to us through land clearing, logging, or burning, contributes to the *gross* simplification of ecosystems, which, in turn, eventually, causes desertification: it should be regarded as an ecological *"crime"*.

That re-establishing forest *will be* a significant step towards helping arrest the increase of carbon dioxide in the atmosphere is supported by carefully studying one of the best, recent records of measured atmospheric carbon dioxide. The measurements resulted from the work of Charles Keeling (1989) and his team based in Hawaii. They showed the rise in atmospheric carbon dioxide to be more-than-exponential over the past forty years. The graph plotting this rise is known internationally as *"the*

Keeling Curve" (you can read his paper via the web). But as much as it is cause for alarm, that very same curve provides us with good reason to hope, for it does not depict an uninterrupted increase: it presents a very regular, predictable, annual oscillation. The oscillation corresponds to the impact of the Northern Hemisphere's boreal forests' setting of new leaves in spring and accruement of fibrous tissue (*evidenced by timbers' annual, tree-ring increments*) throughout the summer. The boreal forests of the Northern Hemisphere differ from those that predominate in the Southern Hemisphere because of the huge number of deciduous tree species. When the leaf-bare, deciduous, forest trees recommence growth with the onset of warmer, Spring, temperatures, they absorb a significant and regular quantity of carbon out of the atmosphere each growing season. This phenomenon **alone** provides us with the confidence that if we replace the world's forest, we can make a hugely significant contribution to reversing the build-up of atmospheric carbon dioxide.

It is a fallacious idea to advocate that short-term, rotational plantations of quick-growing species will suffice for reversing the recent, massive, build-up of atmospheric carbon dioxide. They will not. Achieving ecological sustainability will be aided by recreating permanent, standing forest and providing opportunities for selective, sustainable, mature timber harvesting rather than only planting for short-term paper and toilet-tissue crops. *Permanent* forest creates the stable bank of carbon needed to *permanently* remove excessive carbon dioxide from the air. If a proportion of the mature trees are *sustainably* harvested, we can enjoy the hope that many of the artifacts we create will eventually be as valued as the Tudor chairs and Jacobean boxes now treasured as family heirlooms and representative of carbon sequestration from the fifteenth century. Returning to using sustainably-harvested timber as a staple building material will actually be part of our global solution to climate change.

Thus, we can see that achieving genuine ecological sustainability, in-so-far as it is humanely possible to influence the massive, dare I also say, "*Almighty*", interplay between the geophysical, atmospheric and biophysical perturbations of this planet, means we simply must replant forests to restore and replenish bio-diverse biomass globally. In so doing, we shall accelerate carbon dioxide sequestration and recreate the atmospheric and hydrological conditions amenable to the continuation of, and evolution of, life.

Extending the triple bottom line to a quintuplet of principles

When planning infrastructural developments and public policy to achieve ecological sustainability, we need to be sure we factor some of those, afore-mentioned, concerns into the equation. A "*rule-of-thumb*" guide policy-makers have found useful is "*the triple bottom line*" idea (*Elkington*, 1998), which asks us to consider "*the economic, the social and the environmental*" impacts of each decision made. But when we apply this formula to attempt to determine whether nuclear power or solar power, or wind or geothermal, will be useful for meeting humanity's future energy needs, we actually find such a determination stymied – ***all appear to be equally valid***. But this is incorrect: they are not equally valid. We have a better chance of creating a better "*sustainability heuristic or mnemonic device*" (*terms coined by Dr Brinsmead, pers. com, 2008*) if we extend the triple bottom line measure of sustainability to a "*quintuplet of principles*". If we appraise proposed developments with these five considerations in mind, we can be more confident that we are progressing towards genuine ecological sustainability.

The first extension of the "*triple bottom-line*" criteria requires us to factor into our policy and infrastructure decision-making the proposal's overall energy consumption, **particularly**

appraising *energy-efficiency and energy conservation* **considerations**.

"Energy efficiency" requires that we reduce the potential for waste heat to be produced. Wasted energy often manifests as heat: inefficient devices and applications contribute to the global heat-pollution burden. **Highly efficient energy applications result in minimal waste heat**.

The related idea, *"energy conservation"*, requires us to use the least amount of energy possible to do the requisite task. Regardless of what energy source we harness, striving for efficient energy use and applying energy conservation principles, as advocated by the project, *Earth Hour, 2008*, will help reduce the atmospheric heat burden human industrial and post-industrial activity has caused.

The second consideration relates to some users' ambiguous application of the triple bottom line's term *"environment"*. Often, the term *"environment"* is simplistically interpreted to mean *"natural resources"*. The quintuplet of principles to aid genuine ecological sustainability prefers to divide and replace this term to create two discreet, more particularly qualified elements: the first being *"sustainable natural resource use"*; the second introducing *"the restorative imperative"* which asks us to embrace every possible opportunity for both biodiversity protection and bio-diverse biomass replenishment.

Sustainable natural resource use is the impetus behind the new regulations pertaining to the catch limits and off-seasons now implemented by the Australian Fisheries managers. It is the reason for developing fish farms and for recycling metals and plastics, for converting sewage into fertilizers, and for sustainably harvesting trees in carefully managed forest. The idea of a *"restorative imperative"* is similarly, self-explanatory. It asks of us to repair the damage we have done to the natural world, speedily. This requires a huge organizational commitment from every nation to have any hope of succeeding on the scale now, immediately, required. This imperative is, I think, the single most important reason we have for asking all the world's nations to institute, without reservation, the idea of a *Millennium of Replenishment.*

So, the extended triple bottom line, the *"quintuple bottom line"*, suggests that we consider the impact of the following five criteria when making policy and infrastructural development decisions:

1. **The socio-cultural impacts**;
2. **The economic costs** and implications;
3. **Energy efficiency principles** (*to ensure both conservation and efficiency*);
4. **Sustainable natural resource use** (*not exhausting the resource, always harvesting with a view to allowing the resource to replenish and recycling non-renewable natural resources*); and finally, plan to accommodate
5. **The restorative imperative**: protecting and restoring biodiversity and ensuring bio-diverse biomass replenishment.

If we apply these five dimensions towards determining the appropriateness of the energy-generating alternatives, we shall find that nuclear power has some significant advantages over the so-called *"renewables"* and it is these advantages I shall try to convey.

Socio-cultural considerations pertaining to nuclear power

One of the most significant, and yet over-looked, socio-cultural dimensions pertaining to nuclear power matters is the role communication and education play in enabling a society come to grips with what, in essence, has been a fundamental scientific revolution: the commencement of the *Quantum Era* and the *Nuclear Age*. Only the world's most technologically-advanced nations have been privy to understanding the fundamentals of these scientific insights in their basic educational endeavours. But receiving correct, up-to-date, information is now a critical requisite for all nations for our world to have the hope of achieving genuine ecological sustainability quickly. Promulgating incorrect and misleading information, and even omitting nuclear science material from syllabuses, is irresponsible and it may even be indicative of the effective *"propaganda"* we appear to have been subjected to globally in a series of deliberate maneuvers designed to obscure the true value of the uranium ore resource.

Propaganda

In these times of massive transition, while we move from unsustainable ways of living to new, ecologically sustainable ways of living, huge vested interests will do all they can to perpetuate their organizational infrastructure and revenue streams, even though their activities are harmful to the biosphere.

You are possibly familiar with the idea that there are some people who would stop at nothing to annihilate opposition *"if you stood between them and a bag of money"*. And this is very true. Threaten people's livelihoods and they will believe you to be their enemy. Many money-making organizations have a vested interest *in not sharing* illuminating information: the history of the cigarette industry bears witness to this (*Devra, 2007*). So much money is being made by a few from the prevailing ignorance about nuclear power matters that I feel obliged to warn you that the techniques of the propagandist are alive and well. These techniques are applied in a myriad of ways to thwart fairer global policies emerging in this area. But forewarned will be forearmed and if you can detect and understand the techniques used by the propagandist, those who are benefitting from our collective ignorance on this issue will no longer be allowed unfettered opportunities to exploit and plunder resources from information-poor nations.

Let me make it plain: nuclear power, carefully managed and judiciously and conservatively used, will free us from the need to burn coal and enable us all to have refrigerators, air conditioning, transport (*electric cars for domestic use and hydrogen gas for heavy duty rail, sea and air transport*) and reliable communication infrastructure. With its careful application, and concurrent massive reforestation, we will be able to significantly reduce atmospheric pollution from both transport and coal-fired electric power stations: clear skies in the Northern Hemisphere would be a wonder to behold!

Nuclear power innovation is the right thing to do because we now know how to manage the waste safely (*indeed, genuine waste is almost negligible*) and the alternatives are more costly; more harmful to the natural world; less able to reliably provide energy to power essential industrial, communication and transport infrastructure; or more wasteful of resources and energy.

If your first inclination was to recoil in horror at this suggestion then it is possible that you have fallen prey to "*propaganda*". To avoid this potential for manipulation, we need to be forearmed and know just a little about propaganda and how it can be used to manipulate and distort our thinking.

Basically, the tools of propaganda are applied to further an interest that is not in the recipients' best interest, rather, it is information designed to empower its promulgator/s and disempower its recipients. Information is discredited and becomes "*propaganda*" because it can involve a distortion of truth, be completely false or involve significant omissions.

Propaganda needs to be distinguished from advocacy. I use the term "*advocacy*" in the following fashion. The advocate, like a teacher, or university lecturer, relies on educational conventions. They deal with established fact, seeking to impart a perspective reliant on new, better-informed, knowledge. Facts are established by truthfully building precept upon, accurate, precept. The genuine advocate is also "*disinterested*" in the sense that the material advocated benefits the recipient, not necessarily the advocate. The role of a "*responsible advocate*" is to help communicate established, new truths, more quickly than can generationally-bound educational institutes.

The tools of the propagandist (*and Werkmeister, 1948, provides a good summary*) include: denouncing and discrediting authorities by derogatory references; constructing overly-simplistic scenarios; employing "*purr words*" designed to create loyalty to the protagonist's cause; using the testimonials of individuals who have no authority on the subject; bifurcation; band-wagoning; and association.

Most of those concepts are self-explanatory, but "*bifurcation*" is interesting. It creates "*either/or*" propositions – it polarises the debate by over-simplifying alternatives. In Australia's Darwin City for example, there is a great deal of banter, in the press and amongst political party affiliates, that Darwin is a "*gas*", not a "*nuclear*", city. Well, if Darwin became **both** a gas supplier and nuclear-powered city it, and the Northern Territory, could become one of the wealthiest places in the world. It is not a question of "*either/or*". The Northern Territory need not forgo developing access to gas, but to ensure ecological sustainability, these precious gas reserves should be carefully transported, processed, then judiciously and conservatively marketed to realise their true worth for the irreplaceable purposes they serve. We should not squander them for burning to provide more polluting electricity or transport. But the prevailing ignorance about the nuclear power alternative has resulted in the

Territory's political leaders opting to burn gas and diesel to serve the city's electricity needs rather than entertain the idea of a small, efficient, nuclear power plant.

An aside: in relation to locating the storage and transport loading infrastructure for gas distribution on Darwin's harbour, a resource company's submission to the 2006 Inquiry, *Australia's Uranium* (point 6.218) made the claim that one, ocean-going, tanker alone has the energy equivalent of more than 55 Hiroshima bombs on board. I cannot confirm the truth of this statement, but even if it produced the equivalent of one such bomb, it would appear that a gas plant, and each of those tankers, presents a much greater risk to the public than does a nuclear power plant. Like all tropical places, Darwin has at least two seasonal episodes annually of remarkable, sustained, lightening activity. The Australian Armed Forces explosives expert I spoke to cautioned that a pressure wave emanating from that sort of stored energy igniting, perhaps through a lightning strike to a spill or even a direct hit to a containment vessel, would *"probably flatten everything for a radius of more than thirty kilometres"*. Contrast this sort of hazard to what we know about modern nuclear-powered electricity plants: they have proved at least ten times safer than have LNG facilities over the past four decades; 2,000 people have been killed in LNG transport accidents in OECD countries over the last 30 years but there has never been any accident, breaching or leaking of any container-transported radioactive material (*Australia's Uranium*, 2006, p213 & p334, Tables 6.8 & 6.9). Modern nuclear power plants are designed to be earthquake-proof, lightening–proof and aircraft-crash-proof. One wonders if the vociferous opponents of nuclear power infrastructure development for the Northern Territory can possibly be genuine, or are they simply representative of those who have vested interests in either using natural gas for electricity generation and/or accelerating the export of precious, rare gas and uranium resources from Australia? While the rest of the developing world is planning and engaging contractors to build nuclear installations, Australian political leaders simply do not even mention the idea of nuclear power in their deliberations: too much cowardice and fear and too little understanding prevails.

"Band-wagoning" appeals to our *"flock"* mentality. We feel vulnerable and isolated when we stand alone, so band-wagoning appeals to the security we find in thoughts like *"this is what everyone thinks, so we should think the same"*. We find it is a technique used even more cleverly when we are presented with *"facts"* claiming that both solar and wind power are the fastest growing energy infrastructure developments in the world today: they increase by as much as 50 percent annually. But, as George Blahusiak (*pers. comm. Sept, 2009*) reasons, that a fifty percent annual growth in less than one percent of global energy generation installations amounts to a zero point three or four percent increase in energy provision from these sources. Globally, this is not a particularly significant increase. The 50 percent global growth figure, however, has enormous *"band-wagoning"* appeal but, in reality, it represents *a miniscule* level of support for solar and wind power technology.

Another very effective ploy applied by the propagandist is the use of false associations. *"Association"* in Werkmeister's words, *"attempts to establish a connection, a psychological association, between the idea presented and some object, person, party, cause, or idea, which people either respect, revere, cherish or condemn or repudiate – then transfers that feeling to the new cause, person, idea to either, be cherished"* or loathed (*Werkmeister*, 1948 p84).

Associations that induce *"fear"* are possibly the most effective tools of the propagandist. On the issue of nuclear power, this technique is employed *ad nauseum*. Mention *"nuclear power"* to the Australian general public, or even to visitors from Britain, and it is my experience that about one-third of the people approached will turn their backs on you. *"It is"*, they shudder, *"too dangerous"*. This is the result of, not informed reading on the subject, but effective propaganda that associates anything *"nuclear"* with unavoidable and inevitable personal and public harm. For example, Professor Leslie Kemeny, wrote several years ago: *"The ultimate weapon of the anti-nuclear activist is to try to establish some causal link between low-level radiation and cancer. This false hypothesis forms the centrepiece of most 'anti-nuclear' campaigns. It is a powerful way of frightening people and controlling community attitudes. Many anti-nuclear activists are willing to perpetrate scientific fraud and exercise emotional blackmail to create radiation phobia in the minds of their audiences"* (*Australia's Uranium*, 2006, p 350).

Negative, contrived, fear-filled, associations are particularly effective because fear is cognized in the emotional cortex of our minds. It will trigger *"flight"*, *"fight"* or, particularly in women, *"paralyzing fright"*. It is even more effective if the fear-ridden concepts relate to anything to do with life-threatening events, such as cancer, mutations, radiation sickness, the carnage caused by war and bombs, and so forth. It is effective because such associations cause emotion to overwhelm the cognitive processes and, thus, stifle the propensity we have for rational thought based on truthful information. It is for this reason that we should never make important decisions when we are fearful or when we are overcome with emotion. It is highly probable that such decisions will not provide the wisest course of action.

Western world journalists, with their thirst for *"gripping"* human-interest stories (*indeed creating such stories is almost an economically-compelled requirement of the info-tainment world*) often construct *"news"* to maximize its emotional impact. Such news is regarded as *"good writing"* or *"good dramatization"* for it harnesses the potential for prose, poetry and art to be expressed in our communication *repertoire*. Prosaic, creative, artistic expression attracts and holds the attention of readers, listeners and viewers: it generates a marketable commodity: it sells and makes money for its promulgators.

In their search for Pulitzer or Walkley prize-winning stories, journalists are often the first to feel emotional indignation and outrage at any impost that potentially threatens human well-being. It is highly likely that this well-developed propensity for the emotionally-propelled response of *"righteous indignation"* has significantly clouded the Australian *"free"* press's ability to impartially assess the merits of nuclear power. This great emotional proclivity has made the new-age journalist hugely vulnerable to propaganda that applies contrived, *"fear-filled,"* associations, perhaps even more so than the rest of the population, excluding teachers, mothers and middle-aged women (*the other most vociferous, but least well-informed, groups opposing nuclear power according to extensive surveys published in Cohen, 1990, and mentioned in conference presentations, ANA, 2005, 2007*). Their consequent reportage of nuclear power issues, being the basis of the chain-of-information that creates the powerful news agenda, is then responsible, perhaps inadvertently, for manufacturing *more-effective* propaganda by maximizing the emotional impact of the information they, as nuclear-information-poor-individuals, may have found alarming. Because journalists are predominantly humanities-trained, they may not have had the confidence to properly investigate the nuclear question. They may not have sufficient knowledge to be able to distinguish legitimate argument from spurious rhetoric. Journalists (*and for some time I was one of them – I thought the long half-lives of some isotopes indicated that they posed an incredible hazard to humanity*) have been profoundly responsible for perpetuating the myths relating to nuclear power issues and we, the information recipients, have almost totally succumbed to the consequential, smothering or omission of correct information.

The eight lies discussed in my first letter are all representative of the elements of propaganda employed to dissuade people from considering, let alone adopting, the nuclear power alternative. There are innumerable other examples of distorted information being promulgated. We find documentaries that employ faulty Geiger counters and *"Jaws"* music to background alarmist, one-sided, out-dated information, coupled with visual images taken from unrelated incidents, and so forth, to create fear-inducing associations. We are *"bombarded"* with letter-box-dropped pamphlets like *"Stop the Nuclear Madness: Don't 'waste' the Territory"* associating a waste disposal repository with a bomb. There is the educational website *"Ask an expert"* whose advice advocates wind power, not nuclear power, because nuclear is *"radioactive"*, with the advice provided relying on that word alone to conjure fear and no attempt made to contextualize the hazard; the full page political newspaper advertisements in the Northern Territory's 2008 election campaign that included the insinuation that to dispose of Low Level or Intermediate Level Waste would be hazardous to the health of *all Territorians* when in reality it poses negligible harm, and so forth ....

When one commences to seriously study the nuclear alternative and discovers that most of the commonly held fears and beliefs are not substantiated factually, it is difficult to refrain from drawing the conclusion that this misinformation, especially at its source, must, in part, be deliberately manufactured. If it is not deliberate, perpetuating this mythology is grossly irresponsible and indicative of culpable, sociopolitical negligence.

Correcting misinformation

Conveying the truth of the nuclear story requires that its assessors have, at least, a general science background and the skill that we professional communicators call *"interpretation"* for *"interpretive writing"*. Regarding nuclear power matters, the most responsible communication strategy would look at potential hazards and then consider them in the light of all the advances scientists have made to either, minimize risk, or entirely negate the hazards. Overwhelming fear and the consequent rejection of the nuclear alternative can then be seen to be misplaced and inappropriate because the time-restricted position, that of learning about how to minimize its hazards through *"trial and error"*, has been eclipsed by the meticulous, cumulative work of nuclear scientists from all around the world. That nuclear power production has achieved the phenomenal safety record of *"zero deaths"* (*within any internationally-licensed civil power plant*) in the last eighteen years is a remarkable achievement, a tribute to the dedicated work of nuclear physicists and engineers all over the world. **Nuclear power is a mature technology and used conservatively and cautiously it can be managed to minimize its potential for harm and maximize its potential for good.**

I write all this cautiously. There is so much money at stake, billions and billions of dollars, that it would not be unheard of for an accident to happen somewhere in the world (*heaven forbid*) to ensure that uranium-providing nations remain opposed to nuclear power. Given the incredible safety initiatives now able to be employed to enable nuclear power to be managed without harm to humans or their environment, I believe it would be fair to question whether any incident now was, indeed, an accident, or whether it was a contrived, *evil* tactic to perpetuate opposition to nuclear power. (*In mid 2008, my life was threatened and I was told an accident could be "arranged" to happen in an old reactor if I were to publish these letters, but they misjudged their target and thanks to intelligence loyal to Australia, those who delivered the threat are now looking over their own shoulders.*)

The other social considerations that persuaded me to advocate nuclear power generation for our nation was the need for the entire world to have reliable, high-grade, energy for: a) food production and storage; b) light and power generation; c) all forms of communication; and in the future, d) a reliable, high-grade source of energy for both private and public transportation needs. Foregoing any of these essential commodities would result in a diminishment of all our modern civilizations have accomplished. This latter consideration does not mean that we should not be conscious of our need to reduce energy consumption and dependence. Rather, it recognizes that if someone we loved could be helped by the quick, convenient aid that is possible through modern communication, transport and sophisticated medical technology, and we abandoned these accomplished developments, *"we would regard this as a diminishment of our quality of life"* (Brian Forester, pers. comm. 2005). For these particular reasons nuclear power would seem to be the source of energy that will best enable our manifest civilizations to continue to grow and flourish.

Another reason for my nuclear advocacy pertains to nuclear power's potential contribution to our becoming a genuinely ecologically sustainable world.

Wisely and conservatively used, while it is not a *"green"* technology itself, if nations, such as Australia, were to develop enrichment capability, fuelrod reprocessing, and develop and market the small, useful, mini-reactor power generators and desalinators suitable for isolated regions, and then harvest carbon from coal, instead of burning it, to create the hydrogen economy and sustainable transport industries, **nuclear power energy will provide us with the wealth to enable us to become "green".** The cost of replenishing and restoring the biosphere will be a phenomenally expensive undertaking. It is my conclusion, one commensurate with that made by more than thirty other nations around the world, that nuclear power adoption is correct and, as a corollary development to massive bio-diverse biomass replenishment, it will enable the necessary wealth production needed to pave the way for us to become to become a genuinely *"ecologically-sustainable"* planet.

The economic impact of nuclear power

Not only will nuclear infrastructure development generate tremendous wealth, it is now less costly than the alternatives. Standard nuclear reactors and modular, mini-reactor power generators (*able to be loaded on to a semi-trailer*) have the potential to provide electricity at a fraction of its current cost *if* the infrastructure can be owned and operated by the nation-state for the benefit of the whole nation (*as happens with State-owned electricity-providing utilities, like France's Areva whose quarterly profits this year were more than 4.6 Billion Euros*, in Nuclear Power Daily, 2010).

If the profit-making incentive is allowed to prevail through the privatization of a nation's essential power-generating services, then it is most-likely, their citizens will pay more, even a great deal more for nuclear electricity. I am a strong advocate for the expertise of a nation's best engineers and workers to be employed through their private companies to help design and construct the necessary infrastructure, but I suggest that the responsibility for supplying, rationing, costing and inspecting nuclear power installations, should remain in the hands of government departments where accountability for the service, and its costs to the public, prevail each election day. In this way, isolated regions will not be over-burdened by the actual cost of providing them with electricity because the profits made in concentrated population areas can be used to offset the costs imposed by isolation. This proposition has ecological merit for it will enable the ecological footprint of huge populations to eventually be defrayed through rural re-settlement initiatives. We will not be able to plant trees on the scale now required without some significant programs and employment opportunities being created for isolated, currently desertified, regions.

Nuclear electrical power production is already a cheaper alternative than coal-burning power production in many countries because, firstly, the costs associated with the legacy of harm caused by atmospheric pollution resulting from coal-burning, have never been factored into its overall cost (*Cohen, 1990, Keay, 2005*). Secondly, while the power plant development infrastructure costs are similar for both nuclear and coal, coal power costs continue to increase as a function of the "*sheer quantity*" of the resource involved. The cost of extracting and transporting millions of tonnes to the power plant alone is huge and now spiraling. If uranium is processed and enriched near to where it is mined, as is proposed for Australia by Professor Kemeny, and Hardy in his *Cradle to Grave* concept (2006) for Australia's uranium, there will be significant energy savings through minimizing, what are now, global transport costs associated with producing *yellow cake* (*uranium oxide*) and then sending this to all four corners of the globe for processing into nuclear fuelrods.

In terms of overall resource use, a single nuclear power plant requires only a few tonnes of uranium per year, while a coal-fired power station uses many tens of millions of tonnes of coal (*Comby, 1994, p122*). This provides nuclear with a comparative cost advantage that will always remain because: firstly, weight-for-weight, nuclear fuel has 20,000 times more energy than any fossil fuel; and secondly, the uranium oxide resource can be reprocessed and recombined with plutonium to continue high-grade heat production for thousands of years. With this extension of power production through the use of mixed-oxide fuels, one tonne of uranium is able to provide the energy equivalent of many million tonnes of coal (*Keay*, 2002, p 7 & 2005, p35).

Another factor contributing to the sound economic rationale supporting nuclear power development is that nuclear power generators do not require a duplication of infrastructure as is required to support the geothermal, wind and solar power alternatives. The intermittency of sunlight and wind mean that both of these alternative power sources will always require a "*back-up*", stable, alternative, base-load electricity supply and some energy storage facility, like batteries or heat banks, to draw upon when there is no sunlight or wind. So, instantly, without considering any other factors, wind and solar technology will cost a great deal more than the nuclear alternative. This sole consideration alone makes them uneconomic for the bulk of the world's densely populated areas. The "*renewables*" may be useful in small, isolated, locations for supplying, for example, power to supply residential lighting and support some form of communication infrastructure, but for a wind farm to produce something approximating the electrical output of a single nuclear power plant, tens of thousands of turbines are required. For example, if the French were to replace their nuclear power with wind power, the turbines would stretch one kilometre thick for five hundred kilometres (*Keay*, 2002). And in Denmark, the Danes have the most expensive power in Europe because of their need to continuously import a stable, base-load, back-up supply of electricity (*Supplement in Keay*, 2002). In relation to the solar alternative, not only is it intermittent, most people are not aware that to replace conventional electricity, solar panels will be required in *enormous* quantities. Even worse, solar panel production involves the use of many toxic chemicals: hydrofluoric acid, boron trifluoride, arsenic, cadmium, tellurium and selenium compounds, some of which remain hazardous to life forever (*Cohen*, 1994, p268). These substances must be used with great care, as the following information, gleaned from a quick survey of the internet, reveals.

Hydro-fluoric acid, for example, is very dangerous. It burns deeply into skin and muscle and cannot be flushed out with water. Airborne particles at concentrations of 10 parts per million will irritate our eyes, skin and respiratory tract but at 30 parts per million it is immediately dangerous to life while airborne concentrations above 50 parts per million will be fatal.

Boron trifluoride is toxic at concentrations as low at *five parts per billion*. It is extremely hazardous if it comes into contact with our eyes, skin, or if it is ingested or inhaled. Handling this substance requires face shields, a full-body suit with a vapor respirator and self-contained breathing apparatus.

Arsenic, in any concentrations whatever, inhibits the action of lipoic acid and so, it particularly affects our brains, causing neurological disorders and death. In contact with our skin it can also cause skin cancers. Arsenic cannot be destroyed, it can only change form and, as arsenic trioxide (*mixed with oxygen*), it becomes 500 times more toxic.

Cadmium produces toxic fumes and workers can be exposed to it in the air when it is smelted. Cadmium exposure causes flu-like symptoms, chills, fever, muscle aches and asthma-like lung conditions. Symptoms are noticeable within hours of exposure. Inhalation causes loss of the sense of smell and also causes liver damage and irreversible kidney damage, leading to kidney failure.

Tellurium, is a very rare metal, one of the nine rarest metallic elements on Earth. It is used in blasting caps and was the metal *bonder* used to make the outer-shell of the first atomic bomb. It has eight natural isotopes, and like uranium, three of these are radioactive. Tellurium is used as a light-sensitive-conductor to help increase the energy-conversion efficiency of solar panels.

Selenium is an essential trace element which becomes toxic in large doses, so it too must be cautiously handled and used.

I mention these particular hazards to place the fears associated with nuclear power generation into a better, more-informed, perspective. From the above information, we could easily conclude that it is safer to work in the nuclear power industry than in the solar-panel manufacturing industry. Yet, around the world governments subsidize the extremely hazardous manufacturing process under the guise of their being a natural, renewable · alternative and sustainable energy source ... something to think about when considering relative risks and hazards.

These, so-called *"renewable alternatives"*, wind and solar (*and even geothermal – because another source of energy is required to pump the water below, then more energy is used to super-heat the lukewarm water which resurfaces to enable turbines to turn*), do not really assist us in our aim to achieve genuine ecological sustainability. When we carpet huge tracts of our landscapes with wind towers and solar panels, we deprive nature of the opportunity to process the same energy to enhance life-on-Earth's capacity to restore oxygen and water vapor to the atmosphere and sequestrate carbon dioxide as it is able to do through strategic forestation. These sort of alternative power-generating developments, on the scale needed to support urban living, are potentially inimical to the renewal of the life-sustaining functions of the biosphere: a planet covered in solar panels and wind farms could soon have an atmosphere like that of our lifeless neighbours, Venus and Mars.

Another significant consideration that demonstrates why nuclear power is the best, most cost-effective, solution for our energy needs is the fact that in all past estimates, nuclear power infrastructure costs have been significantly over-stated. Reactors, thought to have a life of perhaps forty years, have been re-licensed, in the United States and elsewhere (*Shandong, Belgium, Taiwan*), to give them at least another twenty operational years. This reduces the infrastructure-provision cost of electricity by approximately one-third and makes all previous, hostile, cost-comparisons obsolete. The costs are likely to remain low because the bulk of the expense associated with nuclear power stations is in the actual building, not, given the very low price uranium oxide has attracted throughout most of its mined history, in the fuel. For example, if the cost of gas doubled, gas electricity production (*such as that proposed for Darwin*) will increase by seventy percent. If the cost of uranium oxide doubles, the cost of nuclear-generated electricity for Australia would only go up by five percent (*I. Hoare-Lacy*, 2005, p3).

A final economic consideration is that if we don't develop nuclear power and we continue to irresponsibly burn coal (*and gas*), we forego the amazing opportunity the world's extensive coal reserves present for supplying the complex carbon molecule feedstock necessary for future generations to be able to make and use plastics, synthetics and polymers. If we squash and liquefy coal, instead of burning it, we can harvest complex carbon molecules and manufacture such materials. Polymers will enable us to proceed with the hydrogen economy and that initiative alone will generate tremendous wealth and employment for many generations to come. It is grossly irresponsible to continue to burn coal when we have had the nuclear option as a feasible, clean, non-polluting alternative for nearly forty years.

The same argument applies to natural gas: burning it is not only contributing to increasing atmospheric carbon dioxide, it is squandering its irreplaceable potential for use in welding and industrial furnaces. Natural gas is a particularly valuable and finite resource. It is believed there will be no more than fifty to seventy years, *in total*, of extraction opportunity for the entire world. It now makes better sense to discreetly apportion its extraction and sale by judiciously applying extraction and export-cost impositions so that generations in future will also have access to it.

STOP PRESS: Angel Publishing's *Energy and Capital* release a story by Nick Hodge describing an 80 percent reduction in costs by using nuclear power generation and desalination, 10th August, 2010.

Energy efficiency issues

I maintain the position of a nuclear power advocate in relation to energy efficiency issues because, while I am not an engineer or a nuclear physicist and it is quite time-consuming for me to assess the merit of all of the hidden assumptions and calculations that underlie all the counter arguments the alternative *"renewables"* attract, I am a former student of some of Australia's best Earth and Life scientists (*including John Chappell, Professor Brown, Ross Florence and Stephen Boyden*). To explain my thoughts in relation to energy efficiency and nuclear power, I draw upon the fundamental ecological insights their teaching provided whilst I read for a science degree at the Australian National University.

Nuclear power has an amazing potential, through the use of super-smooth, low friction turbines, to have tremendous energy conversion efficiency, upwards of sixty percent, even, perhaps in the future, attaining ninety percent plus conversion efficiencies. These sort of energy conversion efficiencies are not possible with the renewables. For example, a solar panel is theoretically only able to convert a maximum of approximately thirty percent of its received energy into something re-useable. In reality, it mostly attains half this sort of conversion efficiency. It is even suggested that to manufacture a solar panel requires more energy than it will ever harness in its entire life, but I cannot measure the truth of this statement. I can, however, make the following observation: we can fry eggs with solar energy but we cannot operate equipment that requires high voltages and consistently high-grade electricity, as is required by industrial refrigeration, air-conditioners or hydraulic lifts, with solar panel power. If we try to, the massive infrastructure required to extract that much energy and then concentrate it, makes the process unviable: it is not energy efficient nor is it resource efficient. Why? The fundamental laws of physics (*Dr Mitroy reminded me of these binding principles*) dictate that you cannot turn a low-grade of energy into a higher grade of energy without expending more energy. This applies to both wind and solar power.

With wind power, a particular current is required for the electricity to be able to be added to the grid. One must convert the direct current into an alternating current. To achieve either requires a conventional, stable, base-load supply of electrical energy to supplement the wind power infrastructure to boost and/or regulate the current continually. Wind power is good *in situ* for grinding grain and for charging batteries in isolated environments (*such as Antarctic base camps where the wind rarely, if ever, stops blowing*), but it will not expeditiously, in terms of energy efficiency or resource efficiency, help to sustain the energy requirements of modern urban living. The turbine maintenance costs are high: with an infrastructure life, limited to perhaps two decades. The natural resource costs in supplying the enormous quantities of concrete, steel, and the highly industrialized materials required for the geographically massive infrastructure needed to countenance either solar or wind power being able to make more than regionally isolated, or localized, contributions to the urbanized world's energy requirements, preclude these options from being economically or ecologically feasible.

Geothermal power is vulnerable to the vagaries of fluxes in radiation and temperatures derived from the molten core of the Earth. It relies on water being pumped to great depths where it can be heated by the retained energy generated by our planet's internal radioactivity. But it is not necessarily a "*safe*" activity. The returning warmed water could potentially bring radioactive pollutants to the surface in an uncontrolled fashion. The process is not able to produce the super-heated steam needed to turn turbines with the efficiency and regularity essential to prevent *"black outs"* or *"brown outs"* in the supply of a base-load, high-grade, electrical current, without the input of additional high-grade energy. Another power source is required from the onset to pump the water the kilometre or two below the surface; then more power is required to heat the warm water and turn it into turbine-turning steam. Thus, the entire process is not energy efficient, but it does produce electricity and that is why it has been thought of as a useful, albeit expensive, power-generating alternative for miners in some geographically-isolated locales. Experimental drilling to great depths is also creating potential fracture points in the stable bedrock of the Archean shield and other stable, sialic crust zones, so there is not a small risk of the geologically stable environment they rely upon, for the heating and extraction of hot water, becoming unstable. Such an eventuality could create an immitigable disaster!

Sustainable natural resource use

Nuclear power can be considered *"renewably sustainable"* in the sense that it can provide a continuous, high-grade, supply of electricity for our entire planet for thousands of years. The advances in processing uranium and re-processing spent fuelrods have resulted in the life of the resource being prolonged by thousands of years with minimal wastage *(Cohen 1990; ANA 2005; Hardy 2006)*. Even when the ore body is used in its entirety, it will be possible to extract uranium from seawater (*if nuclear power is used to desalinate water*), so the supply of uranium is assured even beyond the thousands-of-years-life the concentrated, mineable, ore resource provides. With recycling and mixed oxide fuels, uranium can be regarded as a sustainable, albeit incredibly precious, natural resource. And nuclear power has one tremendous advantage: the energy harvested is high-grade and constant, not a *"feeble"*, discontinuous supply.

Not only is power production through uranium-use, sustainable, applying this technology will enable us to become sustainable in many other activities. For example: I am very concerned that a bio-gas production plant, fuelled by clear-felling local forests, is seriously considered for some Northern Territory communities to help them allay the phenomenal and crippling costs diesel-generated power is now causing for stations and fledgling settlements.

The more suitable alternative is, I think, a small, mini-reactor, power generator that could supply upwards of 10 megawatts of electric power to recharge electric vehicles and enable them to sustainably harvest their forest (*and cultivate sequestrations opportunities, like planting improved timber species, because they are not compelled to clear-fell the entire forest*) *and* electrically harvest the locally prolific gamma grass weed for its cellulose content *and* electrically process the cellulose into paper and tissue *and* desalinate water *and* produce hydrogen gas to power their heavy-duty transport needs.

Such an entire and comprehensive application of energy and resources will be far more constructive, cost effective and energy-efficient than chopping down all the trees to manufacture bio-gas (*a gas that produces more atmospheric carbon dioxide pollution*) to fuel their generator and their vehicles but provide energy for little else. The net result of such a proposition is, I suspect, more harm to the biosphere, not less. In so doing, they would forfeit the opportunity to sequestrate carbon dioxide through sustainable forestry and, in clearing forests for transport fuel, they will exacerbate desertification and the simplification of ecosystems. Australian scientists are working on a mobile power generator prototype; so too are Mitsubishi, Microsoft, Hitachi and Toshiba.

The bio-gas development proposal re-alerts us to our immediate need to assert wise dominion, which brings to mind the absolute importance for all nations to conservatively use this potential abundance of extra-somatic energy and to immediately commence planet-wide ecosystem replenishment: hence, the restorative imperative.

The restorative imperative: protecting biodiversity and ensuring bio-diverse biomass replenishment

The *"restorative imperative"* simply articulates our critical need to replant the world's forests, quickly, to enable ecological sustainability to re-emerge under humanity's stewardship of planet Earth.

To refresh and restate our important, earlier definition: *Ecologically sustainable development requires humanity to minimize damage to, preserve and protect, restore and replenish, bio-diverse biomass to enhance nature's ability to self-perpetuate life through supporting increasingly diverse interrelationships between ever-diversifying, fluctuating communities of co-dependent species. These interdependent communities, together with their legacy of modifying the physical parameters of their local environments, help to create the ecosystem functions that collectively create the conditions we can describe as "a biosphere amenable to the evolution of life".*

I will digress for a moment here because I want to share with you a contemplation about the wisdom of the ancients. It relates to the oldest story known to humans: the story of *The Garden of Eden* and our expulsion from it.

The wisdom of the ancients

In the story, *The Garden of Eden,* we are told that there were two precious trees in the centre of the garden, the *Tree of Life* and the

Tree of Knowledge. They were always to be protected. Humans were particularly forbidden to partake of the fruit of the *Tree of Knowledge* and risked the wrath of God if they did so. The first *"man"* and the first *"woman"* disobeyed and consequently were expelled from the garden.

It is quite probable that we have mistranslated, or even misinterpreted, this ancient allegory or metaphorical, metaphysical, story. This story has not only survived *"the flood"*, it has been translated from its original dialect and re-recorded in many languages including Hebrew, Aramaic, Greek and Latin before the English version made it accessible to us. As a result of this process, it is possible that only some elements of the **original** experience in the Garden remained, and the rest were most probably lost through the passage of time and we have been left to variously interpret the story ever since. In making this interpretation, I am mindful that the spiritual Creator of creation is as equally able to extract a rib and make a human being as He is to create the process of genetic diversity that enabled the human family, Homo sapiens, to manifest differently to its genetic predecessors, *Homo erectus*, the Neanderthal man, *Australopithecus* and other members of the *Hominidae* family. I think it important for the evolution of our spirits and hearts to postulate what some of the missing elements might have been.

The wisdom of our ancient forebears provides us with the scriptural saying, *"First there was the word and the word was with God"*, or we could extrapolate this as *"and the word was good"* because *"all that is good comes from God"*. Perhaps it is possible that the very first word that was uttered from *"Adam's apple"* – Adam's larynx – was in fact *"apple"*. How singularly empowering and revolutionizing the discovery and implementation of language must have been! It would have immediately enabled the development of the human repository of objective knowledge and collective memory. Words are all we have to convert that which is subjective into that which is objective. While smiles, scowls and growls were undoubtedly, and remain, effective methods of communication, they were not as effective as words whose meanings could be unambiguously understood. The *Book of Genesis* tells us that there were other people in the Garden at the same time, but they are not remembered by name. How fitting an accolade to remember as *"The First Man and Woman"* those who initiated communication in this fashion. This interpretation is actually not my original thinking: my maternal grandmother, Sylve, whose family name was *"Cain"*, told me this when I was a little girl. At the same time, she said that Cain was a *"red-headed man, whose family eventually settled from their wandering in Ireland"*. It is only recently that I have thought of her words as being part of our collective, oral history

In postulating the following extension of meaning, I hope I am not detracting from the profound spiritual implications this story has, for in our *"primordial"* state there is no doubt we were in close communion with God. Bonhoeffer, 1927, describes this communion as an intact, spiritual umbilical cord between ourselves and God, the source of all light, life and creation. Christians believe this connection is re-established with an *"in-filling of the Holy Spirit"* – spiritual birth – through an unwavering, personal, inspired, conviction: the simple act of belief in the transcendental, divine, story of Jesus. This is an amazing gift, confounding human understanding, but the *"foolishness of God is wiser than man's wisdom"* (1 Corinthians 1:25). *"Spiritual birth"* is the explanation we offer for a new sense of connectivity with all that which is higher and wiser than ourselves and which manifests *"in spirit"*, the next stage of existence. That we *"fell"* from this *"innate"* state of grace because we succumb to serving our self-interest, whim, lust or inclination, is the temptation, *"the apple"*, that provides evil, man-made and spiritual, with the lever it has always used to harm life.

I was once told a story about spiritual evil: it is like a cunning, vindictive, savage, dog on a long chain. This determinedly, destructive and deceiving, inclination and spiritual force, feigns harmlessness and the innocent and unwary can be lulled to within its reach where they can be, unexpectedly, savaged and scarred. Those who deliberately befriend the dog, *"Evil,"* and play within its ambit, inevitably find themselves *"turned-upon"* viciously, destroyed by what they embraced.

The Reverend Sheets, in part, examines some of spiritual implications of this most ancient of stories and describes *"pre-original sin"* time in his illuminating book *Intercessory Prayer: How God Can Use Your Prayers to Move Heaven and Earth, 1996.* Interestingly, he also notes that the word, *"Adam"*, means, simply, *"man"*, in ancient Hebrew. So, it does not seem far-fetched to me to suggest that this word may have been applied to a particular family group, the family who first commanded language and were genetically distinguished by a missing rib, for several generations, even hundreds of years.

The advent of language would have empowered humans to manipulate their environment with enormous effect. It is through language that our objective understandings are continually refined: they enlarge as our collective and subjective experiences and cognitions develop and grow. We create more words to represent our new insights, and when these words are carefully and unambiguously defined we can translate that which was subjectively thought, sensed and experienced, into the domain of, albeit, negotiated, objectivity. Hence, the *"tree of knowledge"*: as new leaves can eventually become branches, new ideas can sprout new branches of knowledge. Redundant ideas are discarded like a tree discards its dead leaves and branches.

"Apple", while it has a literal meaning, also has a figurative meaning and it may be a euphemism for *"language"*, for just as an apple is the harvest of a tree, the harvest or fruit of language is

knowledge, and vice versa: *"apple"*, the word, being a literal fruit of the figurative *Tree of Knowledge's* continual crops of *"words"*: words representing actions, objects, concepts and emotions. So, it is even possible that we had *"no death"* – the fate of all human being since Adam and Eve - and no *"murder"* – since the sin of Cain - until those particular words came into our human vocabulary. (*There is some scriptural support for this idea, see Romans 7:7*)

However, one wonders why the ancients attest to the *"fruit of the Tree of Knowledge"* being *"forbidden fruit"* and why this was "a *fall"* for humankind instead of the beginning of the great journey of gradual enlightenment? Besides disobeying God, what could have been the essence of this specific, **original sin**?

Perhaps the idea of a *"fall and eviction"* from the Garden of Eden has been mistranslated and they were not evicted at all. Rather, it is highly likely that they broke trust with God's grace and goodness and despoiled their beautiful garden; something we have continued to do with greater or lesser efficacy throughout our entire human history. The original human family's propensity to exploit their surrounding world would have happened even more efficiently because, through Adam and Eve, human beings had learned to talk. It is possible that Adam and Eve not only shared and ate some fruit; rather, because they could describe the exact location of the tree to one another, *they, and their family, ate all the fruit, literally.* Then, they chopped down the tree for firewood. And then, did the same to all the other trees and, before too long, perhaps one generation, they had turned their luscious garden into a barren waste! This selfish activity, indiscriminate-regardless-of-consequences behaviour, may have totally destroyed and created a desert where the Garden of Eden stood (*presumably the "Edom" of the Middle East today?*) and caused them, eventually, to migrate in search of other lands. If this were so, they had failed in the divine injunction to exercise *wise dominion*: they had ignored and failed to execute consequential thinking in partaking of the forbidden fruit of the *Tree of Knowledge* and, in taking all the fruit, they had acted selfishly and disregarded their responsibility to others, and to the garden which required careful tending to ensure its, and their, survival.

God, the source of all wisdom and creative power in the Universe, may have simply let it be known through the revelation of a self-evident truth, that some actions have harmful consequences and are inimical to life and creation, and so He would, of course, have forbade such activity. With the passing of eons of time, it is possible that only shards of the literal essence of this story remained (*the garden, the apple, the original sin, the desecration/expulsion*) and the metaphorical lessons, like our failure to act responsibly because we ignore consequences and plunder and exploit to exhaustion, became lost.

Selfishness is a tendency that *all people* must master continuously on their individual journeys through life for civilization to flourish. People who exorcise this sin have made tremendous advances possible for humanity: the end of slavery; the advent of human rights; rich art, educational and prosaic discourses; and public endowments of great significance. But even so, it is always possible that we will revert to barbarism within one generation, if this selfish disposition and disharmony dominates our relationships. We see this happening when infants are subject to abusive primary relations or when society is flung into survival mode when rendered dysfunctional by wars. Under war's hostile conditions, an entire generation of children can be deprived of the loving and supportive parents needed to provide solid role-models for perpetuating enduring, harmonious, positive, life-affirming, family values. Hitler, for example, an illegitimate, malnourished, paternally-abused, waif, developed a perspective on life that was totally distorted, characterized by strong prejudices and a sense of injustice, which led to gross intolerance. His evil, intolerant, non-empathic disposition appears to have developed and been perpetuated through the pervasive, chaotic, violence, hunger and abuse experienced by many children of his generation as a legacy of the misnamed *"war to end all wars"*, World War I.

Selfishness is overcome through imparting loving, respect-filled and responsible dispositions towards others: the kind of regard that places loyalty to familial principles and to others, ahead of loyalty to one's pernicious self. Loving, supportive, families teach their children how to be humane; how to be urbane. This love is nurtured when children are sensitized to inclusive, empathizing with others through the loving discipline of parents, grandparents, uncles and aunts, attentive to their children's developmental needs. Abusive relations beget a future generation of abusers. We can create sociopaths if we fail to provide loving emotional support to infants. Severe emotional deprivation in the first few months of life can forever *"numb"* that child's ability to *"empathize"* with the feelings of other human beings (*Skinner and Cleese,* 1983). And if we, when little children, are not carefully rebuked and reprimanded for harming others, harm-filled sorts of behaviour can become entrenched in our behaviour repertoires and ricochet, polluting, poisoning and creating discord in other relationships; contaminating all points of contact, just as one bad fruit amongst many will cause the rest to mould.

It may be appropriate to think of the whole God-forbidden-element in the story of the Garden of Eden - taking the apple - as being the cautionary wisdom, *the revelation,* alerting us to the wrongness of thinking, and behaviour, which ignore consequences inimical to the well-being of both the natural world and our fellows. Benny Hinn's (1993, p30) spiritually-illuminating writing suggests *"the garden's apple"* is representative of *"three great temptations"*: the *"lust of the flesh"* (*it being good to eat*); *"the lust of the eyes"* (*we coveted it because it was attractive to behold*); and *"the pride of life"* –

believing our self-will to be supremely important, hence alienating ourselves from God. But if we are God's Creation, we are created for His unique purposes (*Warren, 2002*). His abiding purpose is our spiritual reconciliation with Him, the source of all that impels creation, and then to grow and manifest the fruits of the Holy Spirit, one and all. Partaking of "*apples*" without regard for the consequences to others and our world, is selfish and so is remembered as humanity's original sin. It is a well-entrenched sin. It explains why there are many influential industrial leaders and investors who continue to advocate burning coal and other fossil fuels to maximize their profits, rather than implement safer alternatives. They have even used their wealth to buy the patents that might have provided solutions: "*shelving*" the patents and, in so doing, preventing the rest of the world from benefiting from new, less harmful, innovations (see the book *Internal Combustion* by Edwin Black, 2006). That sort of selfish behaviour is so widespread it has resulted in despoiling the entire planet, not just a single garden. **Had humanity not been distinguished by developing language and not despoiled their garden through selfish behaviour and callous disregard of consequences, it is possible that the human family would have "*lived forever*" in their garden, but our linguistic, knowledge-bound progress with its resultant despoliation,** unless reversed rapidly, **has numbered our days.** However, there is no point in despairing: depression and despondency create inertia. The story of Adam and Eve in the Garden continues by telling us that the *Tree of life* remained and that it would be protected until the end of time.

The wisdom behind this story remnant is amazingly apt for, even today, like the story of the *Tree of Knowledge*, it can be interpreted both figuratively and literally.

Figuratively, this phrase in the King James edition of the Holy Bible "*the Tree of Life remained with God and was to remain protected by Cherubim and flaming sword for all time"* may simply mean that the solution to our plundering misadventure has been, and always will be, there. And it is this: Godly love, that is, respect for consequential thinking that takes into account the well-being of others and the well-being of the natural world, enhances life and lies at the very heart of all that which will allow life to flourish. To love God, humanity must obey His precepts and exercise consequential thinking that results in others and our environment being respected. This same injunction is expressed in social terms by the trans-cultural Golden Rule, *"Do unto others as you would have them do unto you".* In this fashion, life is not only protected, it can be enhanced, culture can thrive, magnificent gardens and great art can be created: under such conditions, the *Tree of Life* thrives.

If we are fearful, suspicious, competitive, backstabbing, sneaking, stealing, abusive, violent, cruel and callous in our relations, the innate beauty of all that is possible for humans to achieve is stifled. If we fail to take account of the effects of our behaviour on the well-being of others, if we abandon sensitivity, generosity, kindness, hope and trust, we will continually build barriers between others and ourselves. Without trust our potential for effective, truthful communication, the sort of communication that empowers, is significantly reduced.

The post-modernist notion that *"greed is good"* is poison to this *Tree of Life*, as has been our overly selfish, callous, exploitation of the natural world.

Being reconciled to an awareness of the absolute majesty and supremacy of God, such as that which happens with the *"in-filling of the Holy Spirit",* as manifests with real belief in the life and *"divine"* teachings of Jesus, enables us, to the extent that is humanly possible, to transcend our personal errors and also provides us with the *"grace"* to forgive those who sin against us. The futility of holding onto harmful, wrong thoughts and actions becomes apparent when we are confronted with even the tiniest sense of Omnipotent Grace. The in-filling of God's spirit, and the subsequent transformation of character the world will witness, is profoundly involved with an abrogation of self when self-centred space is filled with new-found tolerance, patience, joy, gentleness, goodness, self-control, kindness and peace. But there are others who are far better qualified to write on the soul-felt transcendent experiences that happen to help us on this, essentially individual, spiritual adventure. Rick Warren, Mike Murdock and Benny Hinn's books beautifully describe the reconciliation made possible through belief in Jesus Christ and Florence Scovel Shinn's "*Wisdom essays*" provide real keys to scriptural understanding. However, the maturation of our human spirit is, essentially, personally unique, exciting journey. So, let *us* return to considering the ideas embodies by the concept of the "*tree of life*".

Taken literally, the *Tree of Life*, can be equated to the precept that trees, collectively, provide us with the opportunity to have life. Last Century's "*Man of the Trees*", Dr Richard St Barb-Baker, gave us this simple axiom in 1981: "*Planting increasing quantities of trees is the scientific solution to the Earth's environmental dilemma.*" This maxim is amazingly apt given the propensity of trees and forests to influence micro- and macro-climates, to stabilize water-tables, create and improve soils, ameliorate temperature extremes, clean the air, produce food, provide building materials and fabrics, oxygen, fresh water, and at this alarming junction in human history, even arrest and reverse atmospheric carbon dioxide pollution. That the *Tree of Life* remains with God, i.e. *remains with all that is good*, is the very remedy we now need to turn to, to save our planet and ourselves: we simply must replant the forests of the world. The end of the material world we now know is foretold by the ancient prophets in Scripture. Many concerned thinkers believe that "*tipping point*" has been reached *and crossed* and that the havoc wrought by humanities ignorance and selfish orientation is not reversible. But close examination of the quantum world has revealed that behind the most, minute particles of matter there is harmony, amazing order, precision and a degree of self-awareness that has been called "*omnipotent consciousness*": an invisible domain we know as the spiritual realm (see *The Ghost in the Atom*, 1993). This realm's certain existence provides us with several, steadfast hopes to counter the despair our collective mismanagement induces. One of these being the foretold "*spiritual*" rescue of humanity when, through the return of "*the Son of Man*" (Matthew 24:30) - Christ, meaning a return of "*the truth*" - God himself, the spirit of creation, will dwell *in person* with us and all things will be made new (*Revelation, 21:3 & 5*). This may simply be that we recognize the reality that already is: God did not abandon us when we were expelled from the Garden (*soon after he rebuked Cain personally*). He is with us today and we also know that "*where love is, God is*". If peace and harmony, patience, goodness, kindness, self-control - the legacy of love - dominate our relationships, our replenishment activities and restoration intentions, God will indeed have made his home here on Earth. Another hope is the amazing propensity of life *itself* to beget life: "*life's longing for itself*" (*per "Desiderata"*). By humanity helping just a little, by lovingly sprinkling seeds, protecting and nurturing them, abundant life can manifest to make real the vision "*the leaves of the tree were for the healing of the nations*" (*Revelation 22:2*). The third sense of hope is that we shall recognize the essence of our original (*selfish*) sin and commence to overcome it with spiritual maturation and knowledge that recognizes consequences (*the inevitability of Karmic Law*), then work towards overcoming the legacy of damage we have inflicted upon Earth. Florence Scovel Shinn's writing is the best I have discovered on this subject. Correct, responsible ("*response-able*" *as per Bonhoeffer*) thinking will enable us to achieve the necessary, quantum-leap-forward in collective consciousness that lovingly assumes responsibility for repairing the damage we have already done.

Without such hopes, humanity's future, at this point in time, would indeed appear bleak.

To conclude this digression and contemplation of ancient wisdom: the emptying of the *"cups of wrath"* (*from the doom and gloom of the* Book of Revelation, *St John's dream*) may not prevail as foretelling the end of humanity. There are Biblical precedents for this hope. The whole City of *Nineveh*, when warned that their days were numbered by the reluctant prophet, Jonah, repented and changed their ways and they were spared the foretold destruction. We can have hope that even the outpouring of the *"cups of wrath"* can be reversed if we, too, heed the warning to act less selfishly and implement consequential thinking, both, in our relationships with one another, and in our role as custodians of the natural world. This in itself would be a quantum-leap-forward in humanity's collective consciousness that, if achieved, would indeed reflect God's loving spirit dwelling with/in us. Only sublime respect for our own lives, and for all of God's creation, will arrest our potential for creating deserts, and inspire us to replenish the natural world and implement the sensible, sustainable, collective wealth-creating, initiatives we need to reverse our *holus bolus* destruction of the biosphere.

Nationally-owned, not privatized, nuclear-powered infrastructure (*needed to enable us to terminate our dependence on harm-filled, fossil-fuel burning*) will be but one of these life-enabling initiatives.

***

To return to the idea of *"the restorative imperative"*: although re-establishing the forests of the world is often regarded as too enormous a challenge to be thought feasible, this task could be achieved in stages if we consider adopting the *Commoner Principle* and the idea of a *"Tithe for the Natural World"* as

109

helpful additions to our collective, land management planning. The restorative imperative, the last dimension of the quintuple bottom line, may be guided by these two, simple, considerations.

The "*Commoner Principle*"

I coined this term out of respect for the work of the British ecologist, Barry Commoner, author of *The Closing Circle*. It is also a name that reminds us of the wonderful gift of the Commons that Feudal Britain, so long ago, unselfishly bequeathed to its people for posterity. This principle advocates that land managers strategically and judiciously plant one-third of their land to maximize biodiversity and biomass productivity.

Commoner's experience found that, when whole-farm-planning was seriously implemented, it was possible to significantly improve an entire farm's productivity by strategically foresting or planting useful shrubs and trees on approximately one-third of the land area one managed. Strategic planting helped to prevent erosion; reduced wind speeds; provided stock with shelter and shade; caused lambing rates to double and even treble; and diversified and improved potential income streams.

Trees and shrubs provide us with food (*sap, fruit, seeds, leaves*), timber (*cabinet, construction, veneer and fuel-wood*), fragrant oils, honey, spices, silk, fibrous tissue from leaves and bark (*for fabrics and paper manufacture*), flowers and leaves for floristry, and so forth. If we emulate nature's patterns of diversity in our restoration efforts, planting appropriate species on the North, South, East and West sides of slopes and choosing appropriately tolerant species for frost hollows, swamps, dry ridge tops, salty areas, and all the myriad of different localities nature provides, we are likely to create sustainable environments that are, not only utilitarian and sustainable in value, they could also provide food, and afford protection, for wildlife.

If, in arid areas, we were to apply drip irrigation principles and, using a nuclear-powered process, desalinate ocean and artesian water *and* recycle effluent to produce fertilizer for soil-enriching humus to help establish the plantations, we could indeed reverse rampant desertification. In metropolitan and urban areas, city parks and gardens could be reappraised with a view to doubling vegetation content. Nature strips could become noise-absorbing buffers zones and reservoirs of beauty, habitat and even communal food. Verandahs and balconies could become oases of green. In rural areas, private land holders could be assisted with large scale planting on a cost/profit share basis and Governments could afford to pay for this massive reforestation and revegetation effort through such projects' anticipated long term profits and the immediate profits that publicly-owned nuclear infrastructure **will** undoubtedly yield for our nations. Finally, and most wonderfully, because *"trees bring the rain"*, if we were to attempt to replant entire continents strategically, simultaneously providing for well-watered pastures, working within nature's parameters and using the species most suited to each particular location, we will undoubtedly cause local and regional climates to be less arid and hot. Then, in fact, we will be doing more than restoring nature's bio-diverse biomass, we will be enriching nature's own propensity to support the evolution of life on Earth.

The *"tithe"* for the natural world

The idea of a *"tithe"* for the natural world is really a gentle reminder that wildlife (*and native flora*) will now not survive without our carefully planning for its protection through the preservation, and creation of, appropriate, nourishing, habitats. Innumerable species only survive to this day because land managers deliberately protected their homes, even creating artificial homes such as nesting boxes, and ensured that they would have enough food and water to survive. Besides its Biblical antecedent: that ten percent is the minimum we are required to contribute to ensure that there *"be meat in God's house"*, a tithe is the percentile elected as minimum required representation of species and communities for the survival of natural flora and fauna in Australia's national parks and reserves. It seems appropriate that we aim to have this same percentile *"tithed"* to protect wildlife and natural habitat, especially ensuring access to adequate water, on all private and publicly managed land holdings.

Commencing a millennium of replenishment

If we were to use these two principles, the *Commoner Principle* and the idea of a *"tithe for the natural world"*, as the *"rule-of-thumb"* guides for our replenishment efforts, it could be said that the ancient dream of Isaiah, 65:25, will no longer be a figurative vision, but it will become our reality. Isaiah predicted that *"The wolf also shall dwell with the lamb and the leopard shall lie down with the kid; and the calf and the young lion and the fatling together; and a little child shall lead them"*.

Prose interpreted, this simply means: the domesticated and the wild can co-exist, and aesthetic and the utilitarian can be combined through judicious management, and a simple thought (*the little child*) shall lead them, us, to this satisfying, durable, state-of-affairs.

Ozone

Finally, we should consider the other, most significant cause of climate change, the one which is largely ignored in climate change discussions at this point in time: ozone depletion. This is, undoubtedly, a serious concern. The thinning of the ozone layer appears to be most pronounced over the Poles. This phenomenon allows intense radiation to bombard ice molecules and melt them, and so it is another likely culprit contributing to climate change, perhaps the factor that we can be least confident in arresting quickly. However, I both hope and strongly suspect, we will find that even this problem is able to be countered by re-establishing the world's forests. I believe it is no accident or mistake that saw the ancients conceptually place the *"Tree of Life"* close to the Almighty creative force governing our entire Universe (*they said, it "remained with God"*).

Ozone is the amazing, unstable, tri-oxygen molecule that was produced when the first living plants photosynthesized. The culmination of Professor David Brown's Palaeontology lectures at the Australian National University in the late 1970s revealed to his students that the planet's oldest sedimentary rocks' fossil record show us that blue-green algae were the first photosynthesizing life forms and that they evolved under the protection of a layer of seawater surrounding them to absorb the sun's harmful gamma radiation. The record also shows that the oxidation of ferrous (*iron*) compounds coincided with this algae's appearance. Thus, over millions of years, they were responsible for altering the composition of the atmosphere, for one functional by-product of their photosynthesis was the manufacture of atmospheric oxygen. It was the progressive increments to the supply of free, pure, oxygen that gradually accumulated to become a thin layer in the uppermost reaches of the atmosphere, the ozone layer. This layer formed an atmospheric buffer, absorbing intense and harmful radiation, and allowed life to evolve and migrate from the sea to inter-tidal zones and then, eventually, to land.

Pure oxygen in the upper atmosphere forms a dynamic, energy-absorbing, barrier that operates to protect life from the radiation that can damage DNA and cause the cells of plants and animals, without strong exoskeletons, to eventually malfunction or die. Free oxygen, monoxide, and the usual two-oxygen-atoms molecule, dioxide, being very light, will float to the upper reaches of the atmosphere and there, when bombarded with intense, nuclear energy from the sun, these molecules recombine to form the three-part oxygen molecule, ozone.

Ozone forms when a single oxygen atom, energized by radiation, combines with dioxide. The tri-oxide molecule, ozone, also breaks under gamma radiation bombardment and in this way, free oxygen atoms and oxygen molecules (*dioxide*) are continually reforming into ozone, then breaking back into its component parts: single oxygen atoms and/or the double, dioxide, molecule. The continual breakdown and reforming processes result in the oxygen-rich ozone layer absorbing a significant proportion of the sun's radiation. Absorbing energy in the upper atmosphere in this fashion means that intensity of the sun's radiation reaching the Earth's surface can be reduced, so we can say *"the ozone layer reduces the opportunity for harmful radiation to reach the living biosphere"*.

The thinning of the ozone layer is thought to be caused by the injection of pollutants into the highest levels of the atmosphere. Pollutants readily recombine with oxygen and, being heavier and more stable molecules, they stop or diminish the radiation absorption capability of this layer. Ozone breakdown would appear to have been accelerated by the atmospheric nuclear tests of the 1940s, 50s, and 60s, which saw debris, dust and free radicals taken tens of kilometers, straight into the uppermost reaches of the atmosphere, the ozone layer, where they would not normally reach. Chlorofluorocarbons, CFCs, being very light, also float to the upper reaches of the atmosphere and combine readily with oxygen to form more stable and slightly heavier molecules, and so, they too, diminish the radiant energy-absorption function of the ozone layer. That is why they have been banned. Similarly, high-flying aircraft continue to pollute this layer of rarefied air. (*I am not sure that the NASA invention, hydrogen gas-powered aircraft, would similarly pollute this layer, as their only reported waste is water vapor*).

My thoughts on the matter of ozone depletion are this: it might be misplaced to be totally pessimistic about this man-made environmental disaster.

The history of the planet shows that the atmosphere has been subjected to massive disturbances, similar to that caused by atomic bomb testing, in past geological episodes and would appear to have recovered. For example, the Krakatoa explosion was 13,000 times greater than the Little Boy atomic bomb and sent dust and debris for tens of kilometers into the upper atmosphere. But when that cataclysmic event happened, the forests of the world, particularly the tropical forests, were largely intact. It is possible that the atmosphere only recovered because throughout most of the planet's history, since the massive eruptions of the Precambrian and other early geological eras, we have had far less frequent eruptions *and,* in conjunction with blue-green algae thriving and producing oxygen in unpolluted oceans, substantial carboniferous forests absorbing carbon dioxide and release oxygen, and thus, continuously, supply free oxygen to the upper levels of the atmosphere. That we have inadvertently accelerated a return towards pre-Carboniferous atmospheric conditions is alarming because way back then there were no humans on the planet and those conditions would be inimical to most of the life forms we now know.

If we employ the problem/solution dynamic (*Atkinson*, 1990), a philosophical premise that states that the problem and its solutions are part of the same thought construct (*when we can articulate what is wrong, we are in a position to postulate solutions; it is when we do not know what is wrong, we are confounded*), we find the solutions to these afore-mentioned problems are pretty simple. We need to stop polluting and we need to replant the forests of the world. It is possible that a massive, global effort to reforest land, in conjunction with ceasing to pollute both sea and sky, will eventually provide enough free oxygen to begin the process of repairing, what

Gribbon, 1988, called, *The Hole in the Sky,* created by the thinning of the ozone layer. I am not sure that we have any other options, or other hope, the rate of ozone destruction has been so rapid. I also think it important that a concerted scientific effort be made to study the effects of diminished free dioxide, as would be reflected evidentially by the now rapidly changing carbon dioxide/oxygen ratio. Increased carbon dioxide in the atmosphere means diminished free oxygen – and that does not seem to be a healthy state of affairs and may be the most serious, single reason for advocating massive reforestation to commence a global *Millennium of Replenishment,* as an urgent international priority.

Many of the afore-mentioned ideas are closely related to the need for the world to adopt nuclear power *sagaciously.* Accelerating the implementation of a *"nuclear-powered economy"* may allow us to continue with the luxury of air travel: it will certainly allow us to haul freight, and travel the oceans, with a minimal legacy of pollution. This development depends upon either nuclear power *per se* or using hydrogen gas, the production of which must not result in more pollution. Nuclear power's desalination potential has already been mentioned as essential to our vegetation replenishment efforts and we can have confidence that its stable, high grade, reliable, electric current will enable us to design reliable, energy-conserving technological applications that will end forever any dependence we have created on burning fossil fuels.

So it can be seen, that nuclear power development, while not *"green"* technology itself, it is the technology that will enable us to become *"green".*

Conclusion

Nuclear power has a profound contribution to make to our returning to global ecological sustainability. Examining the five elements of *the quintuple bottom line,* as applied in this

116

discourse, results in support for the adoption of nuclear power and shows it to be a wise course of energy generation for all urbanized nations.

If we humans were to invest even a fraction of the wealth and energy we spend on armaments and defense globally, the vision to re-establish the forests of the world would be attainable within a timeframe similar to the time it took to remove them: two hundred years. It will take at least the same level of organization and energy we commit to defense globally to undertake this massive task. There is no alternative course of action for humanity to adopt. We have come perilously close to *totally* desecrating our entire planet, befouling both our air and water and simplifying the web of life to create urban deserts and waste lands, empty of life: once more creating the same sort of deserts that very probably epitomized the conditions that saw us expelled from the Garden of Eden, a garden we made barren. It is time to learn from the consequences of our actions and harvest the fruit of the *Tree of Life* and plant its seed, both figuratively and literally, to exercise, on behalf of all creation, that which is our God-given charter, *"wise dominion"*. Then it is likely that the wisdom of the ancients' antecedent ideal for the human family will prevail:

"He shall judge among many people and rebuke strong nations afar off; and they shall beat their swords into ploughshares, and their spears into pruning hooks: nation will not lift sword against sword, neither shall they learn war anymore" (*King James, Micah 4:3*).

We have cherished this vision for several thousand years, but *there has never been a more challenging, critical, time* for us to make it our global reality, than now.

References

ANA , 2005. *The Facts on Nuclear Science, Uranium and Nuclear Power*, Sydney, Australian Nuclear Association.

Atkinson BMC, 1990. *Reporting Ecological Issues: Steps Towards New News Values*, unpublished Master of Philosophy dissertation, Centre for International Journalism Studies, University of Wales, UK.

Atkinson BMC, 1992. *Ecosystem Simplification,* Nature and Society, Spring: 31, 32, revised 2009, see www.specialistwritingservices.com.au

Atkinson BMC, and Van Der Sommen F, *The Quintuple Bottom Line to Accelerate Ecological Sustainability,* Short paper, 20th International Conference on Informatics for Environmental Protection, Graz, Austria.

Australia's Uranium: Greenhouse Friendly Fuel for an Energy Hungry World, Nov 2006. The Parliament of the Commonwealth of Australia, House of Representatives Standing Committee on Industry and Resources, Chair: The Hon Geoff Prosser, MP.

Bonhoeffer D, 1927, English edition 1996. *Sanctorum Communio: A Theological Study of the Sociology of the Church*, Augsburg Fortress, Minneapolis.

Boyden S, 1990. *Our Biosphere Under Threat,* Oxford University Press.

Boyden S, 1992. *Biohistory: The Interplay Between Human Society and the Biosphere, Past and Present, UNESCO* and Parthenon Press.

Comby B, 1994. *Environmentalists for Nuclear Energy*, TNR Editions.

Commoner B, 1972. *The Closing Circle: Confronting the Environmental Crisis*, London, Cape.

Cohen B, 1990, reprinted 1994. *The Nuclear Energy Option: An Alternative for the 1990s*, 233 Spring Street, New York, London, Sage Publications, Inc.

Ehrlich P, 1981. *Co-evolution and the Biology of Communities* in News that Stayed News, Kleiner, A. and Brand, S. North Point Press.

Elkington, 1998. *The Triple Bottom Line: Sustainability's Accountants,* Chapter four, *Cannibals with Forks in Storer and Frost,* Triple Bottom Line Reporting: its relevance and application to agricultural production and marketing, http:/Muresk.curtin.edu.au/research/otherpublications/75thanni versary/storer.pdf

Florence R, 1979. *Forest Ecology Lectures*, Australian National University.

Future Directions International, October 2005. *Australia's Energy Options.*

Gribbin John, 1988. *The Hole in the Sky, Man's Threat to the Ozone Layer*, Bantam Books.

Hardy CJ, 2006. *A Cradle to Grave Concept for Australia's Uranium,* Four Societies Meeting, Eagle House, Milson's Point, Sydney, 22 February, 2006.

Hinn B, 1993, *Power in the Blood,* Word Books, Nelson Word Ltd, Milton Keyes, England.

Hore-Lacy I, 2005. *Nuclear Power; Current World Status and Future Trends,* Australian Nuclear Association Sixth Conference on Nuclear Science and Engineering in Australia, Sydney, Weston Hotel, Australian Nuclear Association.

Huxley A, 1967. *Propaganda in a Democratic Society,* in Voice of the People, Christenson & McWilliams, New York, London, McGraw Hill.

Keay C, 2002. *Nuclear Energy Gigawatts: Supporting Alternative Energies,* The Enlightenment Press, PO Box 166, Waratah, NSW 2298.

Keay C, 2005. *Nuclear Energy Fallacies: Here are the Facts that Refute Them,* The Enlightenment Press, PO Box 166, Waratah, NSW 2298.

Keeling CD, Bacastow RB, et al 1989. *A three Dimensional Model of Atmospheric CO2 Transport Based on Observed Winds: Analysis of Observational Data* Geophysical Monographs 55: 165 - 236

Kerkoff LV & Lebel L, 2006. *Linking Knowledge and Action for Sustainable Development,* Annual Reviews of Environmental Resources 31: 445 477

Kuhn T, 1968. *The Structure of Scientific Revolutions,* Chicago, Uni of Chicago Press.

Man of the Trees Organization, 2006. *The Man of the Trees: Richard St Barb Baker* http://www.manofthetrees.org See also, *My Life, My Trees* by Richard St Barb Baker.

McGaffin W & Knoll E, 1968. *Anything But the Truth: The Credibility Gap – How the News is Managed in Washington*, New York, GP Putnam's Sons.

McGarry ED, 1967. *The Propaganda Function of Marketing* in Voice of the People, Christenson & McWilliams, New York, London, McGraw Hill.

Murdock, M 1999. *The Law of Recognition* The Wisdom Centre, PO Box 99, Denton, Texas.

Scovel Shinn, Florence, 1989. *The Wisdom of Florence Scoval Shin*, A Fireside Book, Simon and Schuster, Rockefeller Centre, 1230 Avenue of the Americas, New York, New York.

Sheets Rev. Dutch, 1996. *Intercessory Prayer*, Regal, California.

Skinner R & Cleese J, 1986. *Families and How to Survive Them*, Methuen, London.

The Ghost in the Atom, 1993. Editor PCW Davies, Cambridge University Press.

United Nations, *Our Common Future* Oxford University Press in Kerkoff LV & Lebel L, 2006 *Linking Knowledge and Action for Sustainable Development* Annual Reviews of Environmental Resources 31: 445, 477.

Warren R, 2002. *Purpose Driven Life*, Saddleback Church Resources, Zondervan Publishing, USA.

Werkmeister WH, 1948. *An Introduction to Critical Thinking: A Beginner's Text in Logic* Lincoln, Nebraska, Johnsen Publishing Company.

Wikipedia, to source: Geological Episodes; to confirm Feudal origins of "the Commons"; Biblical translations; story Krakatoa.

WRI, World Resources Institute, 2000. *Pilot Analysis of Global Ecosystems Report* Research Topic: Forests and Grasslands, Matthews, Payne, Rohweder & Murray, http://biodiv.wri.org/pubs_content_text.cfm?ContentID=227.2005

International letter
Nuclear power's political implications

My Dear Friends,
The third letter in this series pertained peculiarly to the Australian people and the Australian domestic political scene without a message of particular relevance to a wider readership. Hence, this, a fourth letter: it has been constructed particularly for the international audience and asks that the entire world immediately orchestrate planning for both a *Millennium of Replenishment* and the global adoption of nuclear powered electricity generation.

There is no doubt that nuclear power development, when managed safely and combined with energy consumption reduction strategies, will provide all the inhabitants of our planet with the opportunity to create an incredibly bright, sustainable and prosperous future. More, the deployment of inexpensive, mobile, *"modular"* desalination units, as can be provided by small and mini nuclear electricity utilities, can give us an assured supply of pure water never imagined, or attainable, in *any* previous episode of human history. With these developments we shall be able to judiciously replant the forests of the world, make degraded desert areas green and, eventually, restore natural freshwater stream flows and aquifers. In so doing, we will undoubtedly contribute to the re-establishment of stable, more predictable and reliable, climatic regimes. An abundance of pure water and reliable energy will enable us to replenish biomass and maintain life's abundant biodiversity, and, through massive afforestation and reforestation, facilitate the process of further evolution and the amelioration of climatic extremes.

That these goals are now technologically possible is due to the remarkable achievements of last century's, pioneering, nuclear physicists. One in particular, the Italian, *Enrico* Fermi, *"dreamt"*, envisioned, and understood, the possibility of creating a genuinely sustainable, controlled, chain of *"lento"* nuclear reactions, which could be used to create a sustainable supply of clean, reliable, electricity. The process is simple: through carefully contrived and controlled transmutation, uranium can be turned into plutonium. The plutonium will naturally degrade back into uranium, and when *"burnt"* again (*subject to "slow" neutron bombardment, resulting in fusion and fission energy production*) the same uranium will create more plutonium. Each transmutation, through fission or fusion, releases a great deal of energy, which, when captured, enables a renewable cycle of energy generation (*uranium to plutonium to uranium to plutonium ad infinitum*) suitable for electricity production to be sustained. This sustainable process is *now* remembered as *"Fermi's dream"*. And, although the renewability is not total (*with each cycle a small amount of genuine waste, that which needs to burnt at much higher temperatures than the existing reactors cater for, is generated*) this potential for recycling the metals until every joule of energy is extracted will enable the entire world to leave unsustainable energy-based global economic relations in its historical wake. With this assurance of plenty for all, an entirely new era of economic engagement can now manifest under compassionate, wise leadership, globally.

The renewability of energy production provided by recycling these metals is possible because one isotope of the transuranic, man-made element, plutonium (*"transuranic" means heavier than, with a greater mass than, the heaviest, naturally-occurring element, uranium*), releases neutrons slowly, in the same fashion as does the rare, unstable, uranium isotope, uranium-235. The rare uranium-235 isotope is needed to start the reaction: it is generally surrounded by the stable uranium isotope, uranium-238. (*Uranium-238 is thus called "fertile" and was, for many years, also called "depleted uranium" which enabled it to be thought of as, incorrectly, "waste" material.*) When uranium-238 absorbs two or more neutrons, it can transform into plutonium. Because plutonium is unstable it decays back into uranium, releasing more neutrons in the process, all of which can *transmute* more uranium-238 into plutonium. When tiny, precise quantities of the requisite unstable materials are combined with a small amount of *stable uranium*, neutron capture and neutron emission can be balanced to provide energy for sustainable electricity production for thousands of years from a very, very, small quantity of uranium oxide. The entire world's electricity needs can be met by this process.

Perhaps the most remarkable fact in this process is that the unstable uranium isotope needed to start this chain reaction, less than one percent of the element's material manifestations in any ore body, is a naturally occurring substance. Many scientists are in awe of this fact. They describe the unstable uranium-235 isotope as "*a God-given gift to humanity*".

An assertion of informed, global citizenship (*citizenry committed to genuinely respecting individual differences,* supporting *the manifestation of human rights and the fulfillment of individual potential, including full employment and more equitable resource distribution for all nation states*) is now required to create the international and individual, nation-based, political will to enable us all to confidently assume the opportunities and responsibilities nuclear power generation provides for all our communities. It is the author's hope that by circulating the information in this book we will be provided with another little step towards humanity making "*Fermi's dream*" our reality.

Political considerations

When I first commenced nuclear power advocacy, I contacted our local Environment Centre to share the results of the workshops I had conducted with a combination of engineers, inventors and car mechanics. We had been "*brainstorming*" the alternatives to a fossil-fuel-dependent transport fleet. All options were considered: solar, bi-fuels, bio-gas, hydrogen fuel cells, hydrogen gas, and electric-battery vehicles. Upon reflection, it became apparent that the non-polluting alternatives (*hydrogen gas, hydrogen fuel cells and electric powered vehicles*) required a stable, high-grade, electrical current. For example, running an electrical current through water can produce hydrogen gas: the by-product is oxygen. To recharge electric vehicles at home in one's garage requires a reliable, high-grade, electrical current. But both of the supposedly, non-polluting, *"renewable"* alternatives, solar energy and the wind, provide only low-grade power, intermittently. They also require expensive energy storage facilities and so both options are hugely expensive and not resource-efficient. Conversions are generally much less than 20 percent efficient. These two alternatives will be most useful in grid-isolated applications. In relation to hydrogen gas, if the hydrogen, or the electricity used to produce hydrogen, is produced from burning fossil-fuels, then the use of that hydrogen will also be polluting. If it is produced from nuclear or hydro power, it will be a much, much, cleaner product. So, the study concluded, logically, the only realistic, cost-effective, resource-efficient, high-grade, non-atmosphere-polluting alternatives were hydro-electricity and nuclear power. Most of the world, Australia in particular, has extremely limited opportunities for producing more hydro-electricity, thus, it appeared worthwhile and responsible to examine the nuclear option.

I, tentatively, commenced to do so.

But here, in Australia, the nuclear alternative has been hugely feared and is incorrectly associated with ideas of radiation toxicity and phobias, like its use necessarily resulting in mutations for hundreds of thousands of years. So, the initial presentation of my earliest findings to a local environment advocacy group, findings that supported the immediate implementation of nuclear-powered infrastructure, resulted in an awful, emotional *"stand-off"* with their specialist doctor spokesperson on these matters. I was accused of being in receipt of nuclear industry funding: therefore, whatever I said was *"propaganda"*.

Not true. I have not received funding outside of a Federally-funded, university scholarship which provided me with a small living allowance for part of my candidature. The allowance always required supplementing by my working as a tutor to support myself and my doctoral research. (*Coincidentally, this support ceased after three-and-a-half years, just as I commenced the critical "write-up" stage of my dissertation. There could not have been a more inconvenient time for such support to end. At about the same time, I was advised, "not to waste my time on nuclear matters". At that stage, for all I then knew, the supervisor may have been right, but I persisted in this inquiry because I had an "intuitive hunch" that it was important and responsible to understand this option: preliminary investigations and logic were beginning to indicate that there might not be any other viable, economic and ecologically-sustainable energy alternative for Australia and, indeed, the rest of the world. I was beginning to wonder, why were some nation states pursuing this alternative, while we, here in Australia, were not?*)

With the kind mentoring of a nuclear physicist, Dr Mitroy, I was directed to read Bernard Cohen's *The Nuclear Energy Option*, which, together with Colin Keay's various publications, turned the tide on my own fear and skepticism to the beginning of thinking, then knowing, that the world had commenced to manage nuclear power safely. In fact, the world's most developed nations had been doing so for nearly half-a-century! Worse, Australia had also been a pioneer in this field, but in a most remarkable political turn-about, this research and knowhow was abandoned and stifled for more than thirty years. Dr Hardy, then President of the Australian Nuclear Association, ANA, later President of the Pacific Nuclear Council, invited me to become an associate, later of full-member, and attend various conferences and presentations organized by the ANA. One of these conferences, *Australia's Nuclear Future*, organized a tour of the research facility at Lucas Heights and there I visited the control tower, viewed a reactor and held a nuclear fuelrod in my hands. The fuelrod was an unused one being examined in an effort to provide greater design efficacy. Apart from the need to wash my hands to prevent heavy metal contamination, this was not a harm-inducing exercise. Several brilliant Australian scientists, including the dignified, diligent, John Carlson, then the Director General of Australian Safeguards and Non-proliferation Office, Professor Kemeny and Dr Hardy, answered many of my *"after dinner"* questions and allayed several fears and some misunderstandings. From that point in time I continued my readings on nuclear power matters. Gradually, I became more informed and, then confident, that if the global nuclear science and engineering communities were prepared to advocate nuclear power as a safe and economically viable option, I could confidently continue to access and assess their knowledge and experience. As I have done so, I have found the scientists' representation, that nuclear power is a worthy and non-polluting, mature technology, to be *bona fide*.

The continuance of my investigation eventually resulted in nuclear power advocacy. This was quite a challenging *divertissement* to the established Australian environmental schools that were seriously considering ecological sustainability issues and, as you can imagine, my doctoral thesis' political implications became profound. I was asked to re-write my doctoral thesis twice and its assessment and the final examination results were delayed for months on end. But my peer-reviewed findings, internationally examined, were found to be *"water-tight"* and were complimented, with several *"excellent"* sections manifest. Then came another significant delay. In seeking to share my findings more widely, one pre-eminent publisher stressed that I would have no credibility in the eyes of the media or the Australian public because I was *"not a nuclear scientist*". This was in contrast to the objections of the environmental "*doctor*" who thought nothing I said could be credible because I was obviously immersed in the nuclear science community's pocket! My solution to this last problem has been to have my work strictly peer-reviewed.

In relation to my personal credibility, all I can say is that this book is, what we in professional communication circles call, "*an interpretive, science writing exercise*". As such it seeks to build a bridge from the highly complex conceptual and technical discourse that *is* nuclear science and translate this so that those who are not trained in this area can gain an understanding of some of the fundamentals. This, in itself, is a specialist, communication skill. I am a science graduate from the Australian National University, with training in ecology, forestry, psychology, human sciences and physical geography. I have two first-class strategic communication research degrees: a Master of Philosophy from the International Centre for Journalism Studies, University of Wales (*in this research I pioneered "Ecological Reporting" for journalists*), and now, a doctorate in strategic communication to accelerate our transition to ecological sustainability. I am actually incredibly well-qualified to undertake this communication exercise. However, establishing myself as a credible spokesperson with the media, while this is important, has not been my primary concern. While I hope that this book will go some way towards combating entrenched media antipathy, and re-informing educators, to do that job well requires a significant realignment of the information they receive in their training. Dr Phillip Moore's transcript of evidence to the inquiry, *Australia's Uranium* (2006, p609), supported by Mr Stephen Mann, and also Mr Keith Alder (*a former Commissioner of the Australian Atomic Energy Commission and its General Manager from 1975 to 1982*) in other submissions, noted that "*the most antinuclear people are (along with teachers and middle-aged women) television and news journalists*". Cohen's

(1990) and William's (1998) work also describes similar surveys with the same results in Britain and America.

I have written this particular booklet because I am primarily concerned that the fears the general public have about nuclear power be immediately allayed by substituting misinformation with correct information. Then, hopefully, the world's citizenry will be able to act from a better-informed position and maximise the benefits the wise use of this technology can, undoubtedly, provide.

It is now my considered opinion that the truth about nuclear power has been deliberately *"obfuscated"* for nearly one hundred years, in Australia and elsewhere, to enable *"nuclear-information-rich purchasers of rare earths"* to continue to have unfettered access to acquiring the world's limited supplies of invaluable, naturally-radioactive, materials.

This exploitation represents a very low form of global trade, dominated by exploitive, disrespectful practices, like that which characterised the slave traders' marketeering. I call this manifestation of global, economic, transactions, *"globillization"*.

Can we forgive the exploiters? Yes, if, henceforth, those who have deliberately misled us and exploited their information-poorer cousins, assert the singularly obvious opportunity they have to redeem themselves by creating a new form of **globallization**, one characterised by benevolence and generosity, not exploitation.

We shall require of them to share their technological knowhow, patents and managerial *finesse* to enable all of humanity to take advantage of the wonderful clean-energy production opportunity nuclear powered electricity provides.

From glob*ill*ization to glob*all*ization

Australia is but one of many resource-rich nations that finds itself having to challenge the voracious and exploitative effects of the form of globalization that rides "*rough-shod*" over national autonomy and prosperity. Here, we have been deliberately misled by both trusted allies and by many of our own politicians and their political advisors. There are also public servants, some of whom have infiltrated high-level decision-making circles, who operate solely to support offshore interests. Their actions, and the deliberate, continuous, propagation of misinformation, accumulated *influence* and have resulted in entrenching Australia's abandonment of the nuclear power option for forty years. Over this period, the exploiters have worked tirelessly to acquire as much ownership of Australian resources as possible. They have entrenched a scurrilous, corrupt free-market ideology that has nearly destroyed Australia's independent manufacturing capability and public service infrastructure. Our nation's best economic interests have been overwhelmed by a continual, stream of resource-gobbling *globillizers* and international money-lenders: organizations which were given unfettered access to Australia's resources under the guise of "*free trade*" and a misplaced reverence for the economic power oligopolies and monopolies can wield.

This episode of culpable political neglect is a consequence of most Australian's political naivety, a divisive political system *and* the result of a continually downgraded education system, causing, I think deliberately, contrived, general ignorance about nuclear science.

All countries that have similarly been passively invaded, Greece and Iceland suffering a similar fate, or in the case of the Gulf Wars, and other wars attributable to the developed world's rapacious appetite for oil, *aggressively invaded (it is also now thought "illegally")*, by hungry, resource-poor, globe-*ill*-izers, would concur that particular patterns of engagement *are deliberately devised* to undermine the strengthening of any independent national identity, unity and economic independence in the resource-providing nation. (*Many of the free markets guiding thinkers, and the rationale they have used, is now exposed by an organisation called the Citizens Electoral Council. I reached my own conclusions independently, but have since been impressed by their thorough historically well-researched exposes. I refer to some of their work a little later in this letter.*)

The undermining of sovereign-national economic autonomy is now facilitated by very particular institutionalised arrangements for global trade. On the surface, these private organizations appear to provide an opportunity to cater for all of their participants' primary trade engagement needs, but they actually result in contrived trade situations and the creation of opportunities that, understandably, particularly benefit the organization's instigators: the strongest (*and arguably, the most cleverly-led*) organizations and nations of the world.

Although, in many respects, indisputably, a great deal of worthy development has resulted from international trade arrangements and from the activities of the organizations that engage in global trade, it is because the value of the uranium ore body has been **deliberately obscured**, providing some nations with incomparable, abundant, energy resource security, but none for others, that the world's citizenry is right to demand an end to such "*one-way*" traffic of resources and information.

The exploiters have achieved their enviable position of energy security for thousands of years, not only at the expense of the nations they have exploited, but also at the expense of the world's atmosphere. It has been known for more than forty years that nuclear power would replace fossil-fuel dependency. This fact was very probably realised at the beginning of the 20th Century by the science's founders. But rather than assisting the world to make the transition to nuclear power, the exploiters have quietly continued to acquire more and more uranium oxide for themselves. At the same time, fossil-fuel burning increased and did not decrease, as it could easily have done so, from the 1970s onward. It is because of this unconscionable duplicity that those from whom the ore has been plundered are right to assert their own energy security needs as the principal priority in their future international trade arrangements. They are entitled to ask the exploiters to share the technological developments that will enable all nations of the world to quickly benefit and realise *"Fermi's dream"*. But while this technological transition will help us all to, eventually, achieve *energy sustainability*, our ecological sustainability will only be assured when the world's governing bodies, in their entirety, commit to the ideal of a Millennium of Replenishment and replant the world's bio-diverse forests.

The Australian nation is in the fortunate position of being able to quickly make the transition to nuclear power by virtue of the fact that its social edifices are able to function to advantage its citizenry when they are well-led. But some nations do not have such well-established political organizational acumen. For some near and distant neighbours, it is right and proper that the developed world's future trade engagement and economic relations be orientated to generously *"give"* to help provide the ability of those neighbours to achieve reliable, honest, informed, administrative infrastructure. Then, by further providing to enable affordable, nuclear-powered, energy security, and opportunities for desalinated water production, the world's poorest nations will have the chance, firstly, to have secure food production, and from that vantage point, commence to prosper, which in turn, will allow them to replenish the damaged ecosystems in their locales. It is only in this fashion, by being generous with knowledge, organizational acumen, resources and training, that humanity has the hope of making our world secure and free from disaffected terrorism and greed-inspired wars *and* repair ecosystems on the scale now required.

Confronting globi*ll*ization

Australia has been slow to assume a reliable, pre-eminent position amongst the world's nations on the nuclear science front because, here, the tussle to assert national independence has been confounded by those who are happy that our large, dry, island continent, remain a quarry.

This tussle for national autonomy began properly more than a hundred and fifty years ago when the ideal of an independent, autonomous, Australian nation, a nation unshackled from the Old World and its wars, inspired some truly remarkable Australian orators: Reverend Doctor Lang, Charles Harpur, Daniel Deniehy, William Guthrie Spence (*the founder of the Australian Workers Union*), and other fearless individuals, like George Black and Henry and Louisa Lawson (*who, combined, founded "The Republican" newspaper as described in Barwick's essay, "The Fight for an Australian Republic", Citizens Electoral Council of Australia, 1999, p34. I recommend this booklet for its* **precious selection** *of quotations from some of these remarkable people's nation-forming addresses*).

William Guthrie Spence, a great Australian unionist, tramped hundreds, nay, thousands, of miles across the nation to inform, educate *and* help Australia's neglected, *raggle-taggle* miners, road-builders and shearers rally to establish more secure employment conditions. In so doing, his work enabled a *tremendous* elevation in the living standards of Australia's working men and women. In his speeches, which led to instituting unions, Spence rallied thousands of supporters when he stood tall and said: "*Let us set up a system that the rest of the world will not be slow to follow! (Barwick, 1999, p31, ibid*). And true to his vision, the example and triumphs of the Australian

Workers Union were soon emulated around the rest of the world with the same, profound, impact on people's living conditions. Spence's eloquent rationale for the unshackling of his fellows from miserable servitude and bondage, remains one of the world's greatest, inspiring, speeches:

The aim of 'new unionism' is a grand one, a noble one. The principle underlying and guiding it, is simply the principle laid down by Him who long ago laid the foundation of the great reform - I mean the principle of love for one's fellows ... we all believe in justice, in truth, in honesty. The world today believes in them. The world could not get on at all unless there were reasonable men practicing those great principles. If we are not able to carry them out in their entirety, if we are not able to practice what we preach because of our circumstances in life, we can at least do this much – we can try to change our circumstances by exercising whatever power lies within us, and by so controlling the affairs of life, remove impediments from human progress so that there will be an expansion of the good, of the noble, of the best. All these are qualities to be admired in man, and mark the distinction between higher and lower humanity. ....

Another, beautiful, ideal was flowering to guide the activities of those who sought to create a new and great Southern nation. George Black, a leading Republican, gave dormant, infant, Australia the ideal of not becoming embroiled in the wars of the Old World (*ibid in Barwick, 1999, p36*). This ideal has endured and remains so deeply inculcated in the Australian's persona that Australians have rallied, very quickly, to thrust the Governments that have take us to war, unambiguously, OUT, but not before we, regrettably became engaged, through kinship and treaty alliances, to support cousins in Britain and the United States and Europe. My, and your, generations have the task of repairing the damage done by this embroilment. We can only overcome the legacy of distrust such aggression has created by asserting that which is most noble, generous and compassionate in our future relations with the rest of humanity's family.

Another essay by Noelene Isherwood, entitled, *"The Great Republicans of the 1850s" (p16 – 27, in The Fight for an Australian Republic, 1999, op cit)*, quotes some of the uplifting passages written by the Reverend Doctor John Dunmore Lang. Lang's speeches were formative in warming and rallying the entire Australian people to the idea of creating a new, unified, nation while they were still just a mishmash of colonies shipping all surplus produce out to the Northern Hemisphere.

Here is a brief excerpt from one of his brilliant orations:

"The feeling of nationality ... comes down to us from heaven. It is the gift of God for the welfare and advancement of his creature man.... So far indeed from the feeling of nationality being a mere matter of the imagination, it constitutes a bond of brotherhood of the most influential and salutary character, and forms one of the most powerful principles of virtuous action. Like the mainspring of a watch, it sets the whole machinery in motion. Like the heart, it causes the pulse of life to beat in the farthest extremities of the system. It is the very soul of society, which animates and exalts the whole brotherhood of associated men. And must the young Australian be debarred from the exercise of that generous and manly feeling, of which every rightly constituted mind is conscious, when he exclaims, with deep emotion, "This is my own, my native land!" ...
In one word, nationality, or their entire freedom and independence, is absolutely necessary for the social welfare and political advancement of the Australian Colonies. Give us this, and you give us everything to enable us to become a great and glorious people. Withhold this, and you give us nothing (Lang in Noelene Isherwood, ibid, 1999, p16).

The sentiments expressed by Lang roared and travelled like a wind-whipped wildfire across the continent. The impetus for uniting the people who were then living in the various colonies was impossible to stop. But, as we shall soon see, it was an impetus that was able to be deflected. Australia was not only a source of gold and wool, it, and its near neighbour New Zealand, had become indispensable food and resource bowls for the hungry Northern Hemisphere. The existing, life-giving, wealth-providing, trade arrangements meant that foreign interests would cling, leach-like, to their attempts to quell independent economic activity and autonomy in Australia and do all they could to control supply and suppliers. They would relinquish their colonial controls only with great reluctance and resistance.

Once more, I refer to the publication, *"The Fight for an Australian Republic"*. In Australia's move towards nationhood, heavy-handed and armed resistance, like that demonstrated at the Eureka battle, of necessity grew into more subtle political forms of persuasion and influence. Those seeking independence for Australia used the press and peaceful public forums to rally the nation. They achieved tremendous advancements, with one in particular shining to guide all of humanity: women were given the right to vote. But resistance to Australia becoming independent remained strong, subtle, and unyielding. For example, in the 1890s, when the Labour Movement first achieved genuine electoral representation, a pragmatic royalist, Richard Jebb, devised an ideal he called *"Colonial Nationalism"*, and he advised Imperial Britain: *"Don't antagonise the colonies, or they will do what America did. Give them **almost** all they want, even tariff protection, strong trade unions, etcetera – all with the aim of keeping them onside for what really matters – the connection under the Crown"* (Ibid, Barwick, 1999, p39).

This well-financed idea, Colonial Nationalism, is the one that actually prevailed when the time came for the first Federated Australian Constitution to be created. Since then, Australia has found itself perpetually at odds internally, with policy positions continuously juggled between those supporting national interests to those which uphold offshore investor's interests.

Up until this point in time, Australians have not managed to politically overcome the divisive legacy caused by a Parliamentary system that relies on an entrenched opposition as an integral part of its functioning. This creates a constantly "*divided house*". And, as history constantly teaches us, a divided house will not stand the tests of time. The future political evolution of Australia will, henceforth, draw forth representations and administration that unify the nation, not divide it.

The new political system maintained the vestiges of colonial rule, which, coupled with the adroit manipulations of shrewd financiers, provided the *prelude* to the emergence of the globalized trade arrangements we all experience today.

When carefully orchestrated, global trade engagements can provide for *mutual* economic growth and development. When not carefully managed, they result in exploitation and Australia, and other nations, are used as quarries. The latter scenario epitomises the trade engagements relating to the fate of Australia's uranium resource until now.

Gobbling uranium

The story of Australia's exploitation is one that may prove to eclipse the monumental deception the Trojans experienced when they were "*duped*" by their own beliefs into accepting the villain-hiding, treacherous, wooden horse, false-offering.

The duping of Australia actually started more than a century ago. In 1906, an intrepid prospector found a glowing, little hill in the night-darkened desert, about sixty miles Southwest of Broken

Hill, near the top of South Australia. The peculiar ore body was first named *"Smith's Cornotite Mine"* after its prospector, Arthur John Smith. It was later described as a significant *"Davidite"* deposit and recognised as being rich in radium and uranium and, consequently, named *"Radium Hill"* by Douglas Mawson.

"Radium", while it is an actual element, was also the word then used as the name for a variety of radioactive elements and their daughter, decay isotopes, and it include uranium. The value the market then placed on this ore was very quickly realised. In 1911, Radium reached the amazing price of thirteen-thousand, five-hundred pounds per ounce *(£13,500)*. This is because its potential for energy production *(steam generation)* was realised and known by the literate, enthusiastic amateur and professional scientists of the day. For example, in 1913, an Adelaide newspaper even chronicled the news that *"one ounce of it would be sufficient to drive or propel three of the largest battleships afloat for a period of two thousand years"* and that, for Australia, the ore provided jubilant news: *"it would mean that foreign nations will be obliged to seek from us the power wherewith to heat and light their cities, and find means of defence and offence ..."* *(Kakoschke, KR in "A Clouded History" and Wikipedia, "Radium Hill")*.

But, as history shows, those foreign nations were not having a bar of any such *"obligation"*.

World War I began three years after the Radium Hill Mine commenced operations, and the cream of young Australian men, more than ten percent of the population, were *"conscripted"* and sent to their *certain* death: to Gallipoli and to the trenches of France. This war created huge social dishevelment, causing all of the nation's fledgling enterprises and infrastructure to suffer. The Radium Hill Mine closed, then re-opened after the war. But during the period between its closure and its re-opening in 1921, the knowledge of its ore's value, and its potential for creating unlimited steam power, was lost. This information disappeared from general, erudite awareness: it was drowned by the trauma of the War and the deaths of *"the cream"* of Australia's growing population, many of whom were miners, engineers and amateur scientists.

Was it lost or was it deliberately stolen because those who would have understood its potential - the fittest and most able of Australia's men, those who would have been able to transform potential into reality - had disappeared into the trenches of Europe and sands and bloody sea of Gallipoli? It is a terrible thought, but it begs us to ask the question: to what extent were the competing commercial interests in oil and steam behind the conflicts that drove half of humanity to war in the early part of the twentieth century?

There does exist a small reference to the fact that the First World War was contrived for economic purposes. It appears as a note discussing war as *"the proving ground for national character"*, in the essay on the ANZAC Legend in Dennis *et al's The Oxford Companion to Australian Military History*. This particular patriotic sentiment promulgated the spurious idea that a nation is *"born"* when people are prepared to sacrifice their lives for it. Rubbish. In our case, it was born when we decided to unite under the banner of a Federation of Australian States.

But, prior to that war, there were publications emanating from the Old World about '*a growing belief in the redemptive power of war among European intellectuals, who feared that a prolonged peace produced* **enervation and decadence**' in the population at large (*Dennis* et al, *ibid, P37*).

Such an odd, callous and old-fashioned, idea! The home-grown Australian solution to their young men's *"enervation"* was Aussie Rules Football, and from the mid-1800s onwards, teams were fielded for almost every town and station across the nation. In

Australia today, playing sports remains the same sort of channel for expending "*testosterone-driven*" excess energy in most established families. But another agenda was at play one hundred years ago. We find references that the dominant forces supporting the Empires of the world "*increasingly saw a successful war as the only way to maintain control over liberal and socialist challenges*". Not only that, colonial capitalists wanted to "*secure channels for economic control*" and wars won would give them that control.

Britain then commandeered the world's trade by virtue of her naval superiority, but Germany was determined to challenge their supremacy. That, and other tensions in Europe after decades of "*territorial incursions, acquisitions and forced integrations*", meant the whole European continent was ripe for hostilities. There was genuine fear in some quarters that the world's growing socialist movement, that which had particularly flowered into Australian Unionism, was uniting millions of labourers into effective organisations that would resist wars. These organizations saw war as "*a conspiracy that pitched oppressed workers against each other*". The emergence and strength of this radical notion, and that of Universal Suffrage, was a threat to the various ruling powers, and their response was to both prepare for war and to glorify is as "*a traditional, noble, activity*", one "*recognised as a purgative and refreshing in national terms, an outlet for popular unrest and a stimulus to social cohesion*" (see the *Introduction, The Oxford Companion to Australian Military History,* Pope & Wheal, 1995).

If the first Great World War was deliberately contrived to "*quell and control*" the masses, what economic motives were also present? What role did securing access to rare and incredibly valuable reserves of radium/uranium ore play in that diabolical, contrived, evil, scheme?

The answer is found in that the First World War enabled the establishment of *"emergency wartime powers"*. These powers gave governments and industrialists *"greater jurisdiction over communications, labour and all areas of resource consumption"*, as well as *"practical control over price and distribution of war materials, along with enormous profits"* (*ibid, Pope and Wheal, 1995*).

The Machiavellian plot succeeded. Following the War, the true value of radium/uranium ore remained effectively buried and hidden from the general public's knowledge, in Australia and the rest of the world, from then until now. Its value certainly did not manifest in subsequent contracts of sale. The history of the Radium Hill's mining operations shows that, almost from the outset, and until 1961 when it was closed, *"lock, stock and barrel"*, the mine was plundered to fulfil a contract the South Australian Government signed to supply uranium to a company called, *The United Kingdom and United States Combined Development Agency*.

Another interesting fact is that, before the World War, the developed nations' investigations relating to energy security included both the options of steam technology and fossil-fuel-burning, combustion engine. But by the end of the war, steam shipping, steam cars, steam-driven agricultural equipment and heating systems, all of which would have eventually been able to be powered by harnessing and concentrating the natural decay of the uranium ore, were replaced by petroleum-powered vessels and the combustion engine. Fossil fuel then won the tussle, most probably because it was more easily *"burnt"* than uranium: it was also more readily available and able to be distributed everywhere in barrels. But no other resource in the world can provide the enabling energy that uranium oxide gives. It is actually a priceless commodity. And in the one hundred years that have passed, thousands of tonnes of high grade and easily accessed deposits have been shipped-off abroad from Queensland, the

Northern Territory and South Australia. The bulk of this has been sold through pre-sale contracts that provided the resource for less than $20 a pound! Unknown to most of the Australian population, the 140,000 tonnes of uranium oxide exported from the Northern Territory alone (*the quantity from the rest of Australia, not established*), is actually enough to enable a significant part of Europe to be provided with clean, safe electrical energy for at least a modest 7,000 years *before* recycling.

The value and potential of uranium for electricity and steam production remained known in some quarters – high-level Government circles – but here it became inextricably mixed with the secrecy of pertaining to atomic weapons production. This evil purpose unleashed a reign of sheer bullying terror that we still struggle to overcome diplomatically. Some say that nuclear weaponry enabled the Second World War to *absolutely* finish. I think the best outcome to manifest has been the commitment many scientists then, and subsequently, made to steadfastly and determinedly **only** apply uranium oxide for peaceful purposes. Throughout this time, Australia was privy to the world's nuclear science deliberations by virtue of several, Australian-born, distinguished, nuclear scientists. But just as Australia was about to launch into becoming a nuclear-powered nation, it experienced another renewed episode of cleverly orchestrated, politically-infiltrated activity and national autonomy was significantly undermined once more. The manifestation of this period of distressing and unsettling influence resulted in our abandonment of nuclear electric power development.

In 1969, the Australian Parliament's Federal Cabinet decided to support independent nuclear-power development, but just as it did so, political chaos erupted. National security was threatened by Indonesia's invasion of Timor. The Prime Minister was sacked by the Governor-General. Anti-Vietnam war protests chanted, not "*ban the bomb*", but "*ban uranium*", and nowhere was leadership able to emerge to provide stability of purpose to nuclear electricity development, nor was it able to counter the national unrest and rebellion that characterized the dishevelled social milieu of the day. Instead of quelling misinformation about uranium with education: fear and ignorance and deliberate propaganda about uranium were allowed to prevail: in the media, in political circles and in our educational institutes.

The nation was so effectively hoodwinked on the issue of uranium's potential for generating inexpensive electricity, that soon after the Labor Government consolidated power in the early 1980s, we divested ourselves of Government ownership of the ore bodies and failed to set aside a proportion of the irreplaceable, precious, resource for our own use.

That we, Australians, did not realise the incredible value of uranium until now begs for a moral, legal and commercial review of, not only our own policies and laws, but all the international trade arrangements and business practices that have provided nations like the United States, Britain, Russia, Canada and France with energy security for thousands of years, but left the resource-providing-nation, the non-nuclear-savvy-others, with little or none.

This is a terribly dismaying situation for the native-born *(before and since)* British occupation, to grasp. This form of exploitation was the exact situation brave pioneering individuals wanted to avoid when they fought to have Australia become an autonomous, independent, nation.

Our determination to achieve autonomy has been a constant battle for nearly two hundred years. I could not do sufficient justice to the historic tug-of-war episodes between the assertion of national autonomy and foreign-owned controls by rewriting it: I can only refer you to the republican summary provided by The Citizens Electoral Council of Australia, in the afore-mentioned

The Fight for an Australian Republic and refer you to one example: the history of the Australian banking sector. Described there are some of the details of the battles Australians fought to unshackle themselves from dependence upon an offshore-owned banking system whose interest rates, just before the Great Depression, **very nearly took 70 percent of all the tax revenue raised in Australia!** The accounts provide inspirational reading (see, in particular, the essay on *"Money Power"*, taken from *The Brisbane Worker, January 5th 1907* in Robert Barwick, *"The 1880s and 1890s: The Republican Labor Movement Awakens"*, p40 & 41 in *"The Fight for an Australian Republic" op. cit, 1999).*

Take, for example, when King O'Mallee, an advocate for Australian independence, proposed an Australian-owned national bank to control issuance, reserves, exchange rates and deposits in 1909. In his five-hour-long address to the Australian Parliament, he said:

"We are legislating for the countless multitudes of future generations ... We are in-favour of protecting, not only the manufacturer, but also the man who works for him. We wish to protect the oppressed and the downtrodden of the earth ... Cannot Honourable Members see how important it is that we should have a national banking ... a system that will put us beyond the possibility of going as beggars to the shareholders of private banking corporations? .... However great the natural resources of a nation, however genial its climate, fertile its soil, ingenious and enterprising its citizens, or free its institutions, if its money volume is manipulated by private capitalists for selfish ends, its credit shrinks and prices fall. Its producers and business people must be overwhelmed with bankruptcy, its industries will be paralysed and destitution and poverty prevail...
In the Commonwealth, a National Banking System will so greatly reduce interest rates that useful productions will

increase by leaps and bounds. Wealth, instead of accumulating in the hands of a few, will be distributed among producers. A large proportion employed on relief works, building up cities, will be expanded in cultivating and beautifying the country. National improvements will be made to such an extent, and in perfection, unexampled in the history of the world. Agriculture, manufacture, inventions, science, and the arts will flourish in every part of the nation. Those who are now non-producers will naturally become producers. Products will be owned by those who perform the labour, because the standards of distribution will neatly conform to the natural rights of humanity ..." (King O'Mallee, in Barwick, The 1880s and 1890s: The Republican Labor Movement Awakens, pp41 & 42 in The Fight for an Australian Republic op. cit, 1999.)

And once established the early national bank enabled Australia to achieve tremendous prosperity, national infrastructure development and a significant amount of independence from foreign financial institutions' manipulation. Parliament gave the bank the right to issue the national currency and to create a reserve fund control system.

Much of the early national bank's successes appear to *have had to develop* because the world, hugely disrupted by World War I, became caught up in the Great Depression and then another World War. In the Second World War, Australia was actually abandoned by besieged Britain, yet, having had to defend our own shorelines, we emerged from the fray as a much stronger, self-determining, sovereign, nation. But this period of strength lasted until the late 1960s. Today, the bank O'Mallee inspired, and founded, is now gone. So are the few other, similarly instituted, State Banks: they have all been *"privatized"*. The history of exploitation that happened one hundred years ago has repeated itself. A huge proportion of Australia's taxation and resource revenue, throughout the 1980s and 1990s, and again at the beginning of the 21st Century, has been directed off-shore, either directly or as interest for loan repayments, with recent Labour Governments, State and Federal, borrowing hugely to stave-off the effects of the *"global financial meltdown"* and to finance Sydney's Olympic Games. To prevent exploitation, our borrowings must be borrowings from the future revenue Australian-owned infrastructure will generate. When we borrow from abroad, the profit from the labour of future generations of Australians also goes abroad. This understanding is *"sorry business"* for the financiers to whom the profits now go, but it will be welcome news to enterprising Australians. The huge, privatized banks now dominating the Australian monetary sector often refuse to finance great Australian ideas. Too often they foreclose on mortgage holders at critical junctures, just when an extension of the loan or a when an allayment of repayments would provide breathing space for the entrepreneurs and facilitate the economic viability of their enterprises. While home buyers now receive a modicum of legislative protection from the foreclosure fate, most business do not. Such heartless, contrived practices are some of the reasons for the phenomenal numbers of bankruptcies in Australia. These practices also explains how innumerable, good, Australian ideas have become the purvey of offshore interests.

How easily we have been plundered! Australians have not been vigilant, nor have they been particularly astute. Most of us neither know of, nor are able to read about, the men and women whose hard work bought us the economic freedom and prosperity that once enabled us to lead the world on suffrage issues, on forming unions, on democratic representation, on equal pay, on holiday leave-loading, on providing universal medical care, on providing efficient, inexpensive telecommunication services and other innumerable initiatives. Our schools' history books have omitted to include information about the lives of the Reverend Doctor John Lang, Charles Harpur, King O'Mallee, Daniel Deniehy and William Guthrie Spence. Young Australians now no longer know the story of

William McLean death; they cannot recite the poem of "*How Gilbert died*"; they do not know what George Black did; nor are they aware of the importance of the contributions to the nation made by outstanding men like John Fitzgerald, Anstey, Miller, Lewis and "*jail-bird*" Curtin. As a nation we have neglected Thomas Jefferson salient advice; that which he gave to the entire world: "*the price of freedom is eternal vigilance*". A truism to which we now should add, "*freedom requires of us to **not ever** forget the lessons of our past*".

During the past forty years, when the privatization and dismantling of a huge proportion of the infrastructure, owned and built by previous generations for Australians, has happened, thousands of Australian businesses and thousands of Australian families were rendered dysfunctional and seriously harmed by cripplingly high interest rates and by contrived employment insecurity or unemployment.

When people pour their life's energy into an enterprising dream; or when they use their life's savings to help innovate a new idea, and then lose the lot, it is very difficult to "*rise again*" and recommit to the ideals of enterprise and creativity that first inspired them. The cumulative effect of this sort of experience on tens of thousands of enterprising Australians has helped to cripple and stifle the nation's collective, creative aspirations. We are poorer, indeed, in every sense of the word, because of globalization and our nation's being so effectively "*passively invaded*" and politically undermined.

The tactics of passive invasion

The discussion pertaining to the tactics that appear to have orchestrated Australia's most recent, effective, undermining of national economic autonomy for the past forty years are qualified by this statement: "*For obvious legal reasons, I am asserting and extrapolating "hypothetical" situations. Any actual practices, people and places that appear to be identified by this discussion are 'entirely coincidental and accidental'.* Consequently, I can only mention in barest detail the avenues that *"might"* be currently used to assist another nation to quietly, steadfastly, usurp the nation they are invading from realising policies which are the best and most progressive for its own citizenry. It is also important to bear in mind that Australian political representation has, for more than one-hundred years, been dominated by real division, with *"nary a bridge"* between its two major parties. This division, most recently, has manifest in politicising the public service, with *"jobs for the boys"* being the rationale guiding appointments, rather than assessing the potential of individuals to contribute to national development. So, even without our vast brown land being thought of as a resource reservoir to be gobbled by the rest of the world, (*particularly, and until recently, in Australia's case, the English-speaking world and our former colonial masters*) we have long been inattentive to the ideal and benefits of unified national development. Dreadful sayings are now bandied in the public arena critical of government executive appointments. They describe their witnessing of the resulting fracas of mismanaged portfolios as watching *"pigs at the trough"*: referring to the domination of executive appointments by people who manipulate contracts to favour spouses business interests; whose only interest is to maintain power to continue to provide employment for those who belong to the same club; people who see government jobs as the route to personal wealth and security with no appreciation of their responsibility to generate future wealth and create opportunities for the entire nation. So, it is to our own shame that only a little guile and deliberately enticed division and manipulation has been required by foreigners to enable them to acquire Australia's minerals and resources effectively and inexpensively.

I share my thoughts to alert those who are concerned with building a more equitable world to some of the pitfalls they must avoid on the pathway to creating stronger, self-governing, ecologically- and economically-sustainable nations *and* creating a future characterised by fairer, mutually beneficial, peaceable, nation-building and international relations. Globalization will only be a blessing to all of humanity when "*win/win*" scenarios emerge from the *fracas* of what, too often, has been unfettered, despicable, exploitation and opportunism. That which has happened to Australia, and to many other nations around the world, must be viewed as contributing to, and now concluding, an ugly, immoral, phase in the evolution of global trade engagements. A happier future begs from us all, better forms of globe-traversing, economic engagement, necessarily characterized by more civilised, equitable, caring and cooperative ideals. This potential will not be realised by the unwary. It is only able to be realised by strong, alert, independent, nations seeking to, together, make themselves stronger and happier and hence, create a better world.

Passive invasion

Clandestine infiltrations happen hugely in trade circles. The normal "*buy and sell*", the "*push and pull*", dynamics of applied capital allow those with buying power to acquire the interests they would like to have. For this reason, many successful nations have very strict laws pertaining to foreign ownership: they do not want their own enterprises to become foreign property for that could prevent the "*flow-on*", value-added benefits of indigenous economic activity manifesting locally. Language barriers have assisted Japan, Germany and China particularly in their protectionist policies. But when Australia relaxed her laws restricting foreign ownership, and even created tolerance of dual citizenships allowing overseas acquisitors to hold controlling shares in significant Australian portfolios, including Offices of Parliament, an unparalleled spate of policies, destructive of national economic autonomy, quickly commenced to strip the nation bare. This was achieved by particular, orchestrated policies that oversaw our protectionist policies pertaining to energy, essential service infrastructure and mineral resources, personnel appointment procedures, the education sector and media operations and ownership, being relaxed or abandoned.

How could this significant sort of policy change happen? There is no long term accountability established in Australia's Parliamentary system. A great deal of short-term political opportunism prevails and it appears that policy advice provided by our public service, and other advice taken, and even initiated, by our politicians and their most trusted advisors, was corrupted, placing the interests of globalizers above their responsibility to maintain and provide protection for their own citizenry. Without such protection, it is not possible for a nation to endure as a productive participant in the international economy. One needs to build products and opportunities for local economic engagement by healthy, home-grown research, industry and manufacturing. Then one is able to create products and situations and knowhow that will appeal to the international community's traders. Retrospectively, the governing politicians appear to have been incapable of determining that which was helpful to Australia from that which was exploitative, or they were mere puppets of those determined to plunder, or else they could have been active beneficiaries from the divestment procedures. That it was possible to have inept policy-makers emerge stems from Australia's very open system of political pre-selection processes and a diminishment of wide-representation in political party's memberships (*see the story documented in the 2009 Australian edition of this work*).

The political environment has been a primary infiltration target because it enables other effective "*passive invasion*" portals to be created as conduits for instituting more of the policies that allow the invaders' interests to hold sway.

Let's examine the use of the Public Service, government's bureaucratic administrative arm, as a possible entry edifice. But while we do, I am mindful that I am presenting generalizations. If there is truth in the positions I put forth, then, in future, we must deflect the organizations' activities, from being conduits for espionage, to engagement that provide opportunities for more constructive international engagement.

Actually, in Australia, not much of the Public Service actually remains to provide essential services. It has been replaced by consultants, private, independent "*institutes*" and service providers. Where it does remain it is particularly vulnerable to infiltration through its human resource offices, its personnel departments, and recruitment agencies. It is quite possible that there, passive infiltrators are charged to manipulate appointments.

In relation to having been able to infiltrate and undermine the strengthening of Australia's nation-building capacity, I imagine the process to be quite simple. Starting in a small, innocuous and unassuming way, in the search for suitable personnel *and* encouraged by the notion of "*outsourcing*" to minimise the expense of having to support a very sophisticated personnel department, recruitment agencies, many of which are international, are hired by government departments to appoint "*the best person for the job*". They may even be given the task of determining the qualities and activities of future employees when they are hired to draft "*position descriptions*". These agencies are then in a prime position to discard Australian applicants in favour of international appointees. Initially, the appointees need not even be made at a high level. For example, they can be secretaries recruited from abroad (*not difficult appointments to manipulate*) but when the secretaries are appointed to serve the highest authorities in the land, such appointees can be privy to every new initiative and every important decision being made in various offices across the nation. If not a secretary, the appointee could be an editor, an information technology specialist or a communications manager. Such low-level appointments are chosen because it is easy to convince mangers that no particular "*local knowledge*" is necessary to perform these functions, just good organizational ability, a knowledge of computers, a fast typing-speed or sound editing skills.

When foreign appointments commence to usurp talented local appointments, a steady stream of useful intelligence is able to be ferried around the world by the invaders. Marvellous espionage networks flourish and information exchanges happen through clubs and affiliations that operate openly all over Australia. These networks are able to manipulate for subtle repositioning of policy through indirect, or directly-contrived, intervention. We policy-makers and editors know that documents can have their entire content and meaning undermined by an altered sentence or even by a single, changed, definition.

Take this actual case in point. While that which follows may, or may not be, an actual instance of passive invasion, but it does show how easy it is for a single, misspelt, word to change the entire meaning of a *communiqué*.

In an important publication summarising Australia's knowledge in relation to uranium and nuclear power, the definition of "*mixed oxide fuels*" was the only editorial "*glitch*" in the entire publication. This particular error served to hide and obfuscate the true value of the ore resource. It prevented the reader from understanding that "*mixed oxide fuel*" enable the use of *the entire uranium ore body (particularly the 95 percent of the ore body hitherto described as depleted ore or "waste") for energy production.* In this particular publication, one designed to inform Parliamentarians themselves of the uranium/nuclear story, "*mixed-oxide fuels*" were represented as "*A fuel fabricated from plutonium and depleted **ora** natural uranium oxide which can be used in standard light water reactors*".

Clever, eh? Diabolically so!

The definition provided in this important book is nonsensical. It fails to alert the reader to the value and longevity of this fuel source. It should have read something like: "*A fuel fabricated from plutonium mixed with either depleted uranium oxide, or a natural uranium oxide, which can be used in standard light water reactors and is able to prolong energy production from the entire uranium ore body for thousands of years.*"

The incorrect and misleading definition, the only error in an entire 700 page publication, would have happened in, or should only have endured until, the final production stages, where all such errors would normally be rectified. The only possible conclusion to draw is that it was deliberately contrived. That it happened is, I think, indicative of the extent to which a country can be infiltrated through *varied techniques and in a huge array of positions at every level,* by people not working in its own citizenry's best interest.

Imagine what could happen when *an entire institution* is given over to recruitment that favours the novelty of loyal-to-another-nation appointees, instead of those who have genuine cultural and historical community-based-local affiliations, *and disguises this by espousing "globalization" and "multiculturalism"* as its operational philosophy: it will have become a passive invasion **"entry portal".** And Australians did not know, nor did they imagine, this could possibly happen. The generous, rather unworldly, Australian persona has no real guile. As a people, we have been too isolated and well-fed. We have been caught unaware by a hugely contrived intentional and continuous invasion: one aimed at strengthening Old and Other World interests with no interest in strengthening our independent, indigenous, national development. Please particularly note that this essay is not anti-global trade; it is simply anti-global exploitative trade. Win/win trade scenarios must now be instituted to prevail to nurture developments able to provide better quality of life everywhere on Planet Earth: this can only happen if those who have been exploited assert a determination not to remain so vulnerable, unworldly and gullible.

Entry portals

The local organisations possibly used to assist passive invaders in undermining Australian national-identity, include, not only political parties, recruitment agencies and private institutions, they include: port authorities; educational institutes; media organizations; IT spy-ware providing companies; and, international law and accounting firms whose peak invasive achievements must include law changes that include the policies of *"dual citizenship"*, *"privatization"*, *"centralization"* and *"deregulation"*.

Ports: they are able to become *"entry portals"* because, under conditions allowing passive invasion, direct, accountable, government infrastructure control is replaced by private management and the company need not even be indigenously-manned and owned. In Australia, the relaxation of local port control has been exacerbated over a period of several decades by the fact that, a) maritime training institutes have been down-graded, then found unsustainable, then abolished; and b) because Australia's own merchant navy and its merchant fleet were deliberately dismantled. Where once we sailed more than one-hundred-and-twenty ships under the Australian Merchant Navy flag, we now have only a mere handful of ships (*the bulk of the vessels sold and now trading for foreign companies*). I concede that *it is possible* that our entire fleet had become technologically redundant because nuclear powered vessels are much more efficient and cheaper to run than ships burning fossil fuels (*they require re-fuelling once in thirty years and can produce all the freshwater they need to make life comfortable of* *board throughout their voyaging*). But the responsible course of action would have been to build a technologically-relevant fleet, not abandon our autonomous shipping. (*This is another small example of where we Australians have, through our local opposition to nuclear power, been our own worst enemies!*) One implication of a divestment in national shipping is that the terms of trade for the bulk handlers, thereafter, are no longer governed by what is in the best interests of Australian industry and employment. They are governed by the best interests of offshore shipping interests and all procedures, from cargo handling to watch-duties and wages, henceforth, operate to foreign standards. Terms of trade and shipping standards can then be manipulated to compliment opportunities for enhanced commercial exploitation of Australia through the concurrent, contrived manipulation of seasonally, variable, floated, currency values. It is also possible that where a Government has divested its control of ports, minimal bureaucratic and customs scrutiny happens thereafter. Furthermore, if a privatized port authority becomes corrupted, illegal immigrants can enter a country with less likelihood of detection.

The failure to be vigilant can also lead to the nation being flooded by another potent weapon used by passive invaders: drugs.

Australia has spent more than three decades combating *floods* of various harmful, addictive, illegal, drugs; a great deal of which finds its way into the country through poorly supervised seaside gateways and island-hopping, private, light aircraft. A subtle, tacit, tolerance for these substances is created by the media's portrayal of people embroiled in this undermining activity. Though the attention may be condemnatory, the mere fact of providing visual images of drug-taking and -dealing serves the purpose of legitimising the activity in people's minds. This is one of the little understood *"passive"* but significant effects of the modern multi-media-communicated messages.

The deployment of drugs is not a new tactic: it has been used for hundreds, perhaps thousands of years, to contrive situations favourable to unscrupulous manipulators. Drugs like cocaine, heroin, and *"party drugs"*, including *"ice"*, *"ecstasy"* and *"speed"*, are almost as effective as bullets at terminating sound–minded contributions to citizenry. They take a little longer to terminate life than do bullets, but they have the added destructive impact of furthering crime, making neighbourhoods unsafe, destroying conviviality and helping families to disintegrate. They cost an absolute fortune to police, with little hope of successfully combating the problem once addiction has developed. They are certain *"kill joys"* when it comes to strengthening national identity.

Educational institutes can also be used as portals for passive invasion.

In education there are two diametrically opposed forces at play. The first is the imperative for free and unfettered, accurate exchanges of information and ideas for a discipline to flourish and be useful to humanity. The other is the force is that which seeks to curtail, control and monopolize information for the purpose of political and economic power. The former force has, unfortunately, gained the upper hand in Australia in recent years.

Australians, historically, have not been confident in their own educational institutions and not being able to compare standards easily say, for example, between our own colleges and the Oxfords and Harvards of the world, resulted in long-standing awe. We need not have been so awe-struck. In this important arena, education, we assumed world best status through the work of the Australian National University. This national positioning has been eroded over the past thirty years. And while there are many genuine foreign-born academics working hard to enable Australian institutes to flourish; there are some who are not.

Speculation suggests that the goal of the "*globe-gobblers*" is likely to be threefold: a) to recruit the influence of more invaders through hijacking teaching, research and administrative appointments; b) to brain-drain the best of native-born intelligence by preventing local academic recruitment, thus causing locals to "*go abroad*" for interesting, rewarding, careers suited to their areas of expertise (*the fate of the Australian nuclear science community, in particular, could epitomise this tactic's effectiveness*); and, more diabolically, the passively invading appointees can, c) deliberately corrupt local educational initiatives, those that might strengthen national identity or elevate national educational standards. This would have the "*double-edged-sword*" effect of making off-shore education more appealing and Australia a far less appealing destination for the highly competitive foreign student market.

There is evidence of remarkably poor policy manifesting in this area. It is difficult to understand why the school-based chemistry and physics curriculum must be abandoned, and entirely new schools of thought taught, when one reaches university. Curricula to enhance national development would require us to teach the more-relevant, recent understandings from primary school-level onwards. Similarly, where did the idea to abandon teaching the formal rules of grammar come from? This important discipline was replaced by tolerance for misspelt words and poor grammar and results in the loss of attention to subtle differences in meaning. It is difficult to countenance how such damaging and downgrading positions developed and were entrenched for decades without the explanation that "*we have been deliberately undermined*" being thought appropriate.

Perhaps one reason for this was that the continuity of appointment needed to thoroughly appraise and continually improve education standards was severely hurt when a policy abolishing tenure for Australian Public Service and academic appointments was instituted and a large proportion of the intellect-nurturing workforce was made casual, or worse, redundant. Such policies, providing no security of tenure to public servants, have orchestrated more than three decades of constant, disruptive change. The entire service has become a hugely demoralised and *"frightened-for-their-jobs"* body of people (*more of this story is found in the Australian editions*).

The undermining continues, even today. Policy advice provided to, and enacted by, Australian Members of Parliament has recently prioritized building new infrastructure, school assembly halls built of tin, in all Australian primary and secondary schools. This innovation follows several decades of severe funding cut-backs, the introduction of exorbitant fees for tertiary studies, and the determination to curtail and centrally, politically control the academic disciplines researched and taught at various Australian Universities.

The expensive, decorative, building policy epitomises how effectively this particular sector has been *"invaded"*. If one seeks to upgrade national education, one must commence by improving the training teachers receive at university. Building tin assembly halls will simply not do the trick.

Similarly, a real commitment to improving education requires the academy to be supported and its intellectuals allowed to pursue, with creative vigour, *the passionate*, productive enquiry love-of-a-discipline fosters. By replacing this passion with politically-determined, economically-rationalized *"utilitarian"* educational agenda's, the intellectual health of the nation has been dealt a savage blow. Entire disciplines and brilliant academics have been made redundant.

Universities throughout Australia have been severely under-resourced and even asked to assume an impossible-to-achieve, self-financing, *modus operandi*. In stark contrast, the rest of the developed world continues to provide increasing support to nurture the best possible tertiary educational opportunities for their populations in recognition of education's indispensable contribution to national development. Instead of supporting education and making it inexpensively available to all Australians, exorbitant fees now apply for every course and an opportunity to fill universities with interested life-long learners (*whose enrolments would contribute to maintaining viable class sizes*) is seriously hurt. With this demise much *joie de vie* has disappeared from our lives: opportunities for studying music, fine art, literature, textures, ceramics, mechanical engineering and photography, to provide worthy life-long hobbies and facilitate enriched cultural development, are now prohibitively expensive and universities are empty of students, providing more cause for their staff to be downsized and made redundant. Through such policies, a vicious, self-defeating, cycle of endemic, socio-cultural demise has been instituted.

In part, some of these policies result directly from the Liberal/Labour divide that is instituted in Australian politics. The Labor Government is inclined to support trades people and as an historically orchestrated movement, disinclined to neither understand, nor support, intellectuals. The downgrading that has happened to Australian education is one very strong reason for us to end the political process that foists the role of opposition onto half of its elected members instead of a process able to nurture every member of parliament's ability to serve their electorate's needs and develop each electorate's potential by empowering them with financial resources and the intellectual and organizational support of the entire Australian Public Service. (*The remedy to this historical situation will not be difficult: it need not even, initially, require a change in the Constitution, just a slightly different, round-table, seating arrangement in the Parliament and a preparedness of the members to listen to, critically appraise the merit of, and provide support for, the development needs of each Australian electorate. This idea is discussed further in the essay advocating an "Independent Alliance for Australia").*

Another awful result of successful passive invasion in the Australian education sector is that it has been possible for *"third class"* foreign academics to be appointed to *"first class"* roles. As a consequence, Australia's educational initiatives could devolve to even lesser standards. This is a certain recipe for national disaster! Each and every nation's intellectual aptitude is determined by *the calibre* of its leading thinkers *and* the organisational arrangements they create for generating and sharing knowledge. Undermine the manifestation of, and nurturing of, excellence here and you will most certainly undermine the best of what a nation can achieve.

Passive invasion in education is likely to be evidenced where the predominant new appointees are foreign-born and stay briefly, not even completing their contracted term of engagement, before moving onto other, less-readily, infiltrated organisations for more permanent engagements. This *"yo-yoing"* pattern of appointment creates another vacancy likely to be filled from off-shore and a cycle of discontinuous staff appointments is perpetuated, disrupting the academy and intellectual continuity of thought, destroying the course content of degrees and, simultaneously, allowing the tentacles of foreign *"espionage"* to be extended to influence other Australian institutes.

The trauma experienced by the Australian education sector has been hugely destabilising for our nation. An academic institute is not made of bricks and mortar: it is cultivated from the ideas and knowledge gifted people are able to impart to each new generation of scholars. As 'such, it resides in the established community of scholars, *the academe*, and not in a building. The Australian *academe* is not only undermined by offshore opportunism. It is undermined by the casual employment policy which was *"coincidentally"* legislated for in this same period.

Too many gifted Australian pioneers of knowledge have been given short term appointments to devise new, relevant courses, which are then quickly adapted for computerised teaching, but then, the academic's personal services are made redundant. The course materials they developed can then be handed-over to less-costly-to-employ, less-qualified, permanent, appointees: people whose knowledge base has not encompassed all requisite fields and thereafter precludes them from being able to answer a bright student's questions. Students are disadvantaged and whole schools of thought gradually disintegrate. Uncritically-reproduced teaching notes gradually accumulate nonsensical errors and, over time, the relevance of that body of knowledge fails to be adapted to address the problems of the day and, so, becomes increasingly irrelevant. Not only have university's been affected: Australian's excellent Technical and Further Education (TAFE) system, *and* its apprenticeship training system, were almost entirely destroyed by similarly contrived malpractices and policies.

Another institute sabotaged by policies supportive of passive invasion: the media. It is obvious that once a nation's media is largely owned and controlled by offshore shareholders, its agenda will be subtly modified to protect the new investors' own interests, not those of the Australian national family. For instance, if 70 percent of a company's shareholders are investors from the Northern Hemisphere, it is probable that *"news"* which challenges the economics and sustainability of more investment by those same players, such as the construction of an LNG facility or a tidal or wind farm, may not be presented as *"news"*. It is possible that the nuclear alternative, which will provide a far more sustainable and reliable energy source, could be totally neglected, or even portrayed as an *"unconventional and unsafe alternative"*? It is possible that the prevailing news agenda will repeat global content and neglect important nation stories? Just possible, eh?

Most Australia's print media is now significantly foreign-owned. Even the Australian Broadcasting Commission has commenced the institutional pattern that precedes "*privatization*": it has been "*corporatized*" and made accountable for generating revenue to support its running costs - an impossible task without canvassing advertising or sponsorship. Continued, cost strictures have, for several decades, prevented many talented employees from exercising their program-making skills, precluding a substantial opportunity for the international sales of their wares. The entire corporation now seems poised, on the brink, but a-step-away from being made available for private and foreign ownership through stock-market listing: a procedure that normally happens when an Australian-government-commissioned and administered organization is turned into a Corporation. Disaster: then more of the messages that could help consolidate Australia's national identity will be whittled away to platitudes endorsing the only product brands that can afford televised advertising. These are not locally-made products, but those produced by multinationals. This phenomenon can be called, forgive the play on words "**c(h)oka-culture**": other have called this "*cola-culture*" or "*coka-culture*". The eventual effects of abandoning Australian ownership and control of its national media facilitates a further collapse of the local, critical, scrutiny of a Government's policies, other independent thought and reporting, and, through cost-rationalized strictures, a dearth of local dramatic productions and documentaries. For example: thirty-years ago, "*stringer*" correspondents existed to support the work of the Australian Broadcasting Commission in almost every town in the nation. They are there no longer. That representation is gone. With it, Australia's ability to create and reinforce an independent national identity for herself is also largely curtailed. In its place, we have as our news of the world, the "*mediated*" message's created by the huge multi-media cartels. Fergie-like flaps drown scant attention to the corruption of the privately-owned currency coiners: do we detect a conflict of interest emerging in such agendas? Yes. Australia's interests are under-represented and under siege.

There are other concerns. Television and video watching have largely replaced traditional, community-based communication opportunities (*after-supper musical and debate societies, soap box speakers, town hall meetings and political assemblies*), resulting in diminished, face-to-face political engagement, diminished political awareness and scrutiny and diminished local political accountability. TV/video-culture has also helped to encourage the last three generations of Australians to abandon being "*reading*" people, and arguably, less-critical-thinkers. This is possibly because the wide-spread practice of watching televised imagery lulls our minds into the uncritical, but highly receptive "*alpha*" state of mental activity, reducing people's opportunities for reflective, in-depth, thinking on important issues.   Reading and writing, with particular attention to reasonable, logical assemblages, sharing oral histories and creating opportunities for person-to-person dialogue, are critical to the development and evolution of strong analytical and reflective intellects.   Computer-mediated messages can be saturated with misinformation and vision-based, brief, "*news blips*" of information do not provide for this considered, form of, intellect-nourishing, dialogue. All-in-all, 'tis a rather sorry state-of-affairs.

We, Australians, need to give serious consideration to redressing the dearth of nation-strengthening, media and news agendas. We must find ways of encouraging "*reading*" as a hobby and recreate opportunities for "*face-to-face*", local, political engagement and discussion to enable successful, nation-strengthening political activity to flourish as it has in our past.

Information technology: I do not want to go near the potential security interfaces and computer programs developed abroad have for facilitating espionage, just need to point out that wherever telecommunications allow direct or remote access, all of that computer's contents can be, and is, accessed by determined parties.

Law and accounting firms are other organizations easily used by infiltrators. Traditionally regarded as *"the most insular of professions"*, Scholte and Robertson, 2007, *(who have assembled a four-volume compendium examining all the various dimensions of globalization, see References)* think these organizations have been hugely instrumental in promulgating and entrenching the exploitative forms of globalization we now know. Here, the infiltrating, multi- or trans-national's, target has been to organize legal relationships between business activity and the law *(local, national and international)* to eventually allow foreign or multi-nations interests to usurp, and plunder, another's *national* interests. Who is in a better position to determine the economic potential of an organization than the accountants who regularly appraise its activities? The plundering happens by stealth and under the guise of each nation being persuaded that they have an imperative to support the *"global economy"*.

Combine this with the impact of the passive invaders' globalizing laws and, eventually, it is singularly possible to cause an undermining in the wealth potential of, and integrity of, the economies with which the trans-nationals engage by enabling clear pathways of *"unopposed commercial activity"*, and ownership opportunities for monopoly cartels to swallow and swamp local endeavours.

"One World Government" appears to be the organizational idea that might satiate these super-financial powers' voracious appetites for monopolising the world's economic activity. But *"one world government"* provides little, or no, local accountability. It means that that which is in the best interests of particular communities will very likely be eclipsed by the

interests of those who control purse-strings from afar: those who will wield the economic-decision-making clout. One World Government is a fairly horrific, regressive, idea which could easily provide for the entrenchment of ruthless, inextricable, economic slavery and exploitation, not just for particular groups of people, but for entire nations. Remember, 70 percent of all of Australia's taxation revenue was being paid to such purse-string controllers in the early days of last century. That is economic slavery. No nation in the world should be subject to such extortive, financial, plundering.

The Scholte and Robertson *Globalization* assessment provides several essays that are highly critical of the world's economic relations. They include an assessment of how the activities of the World Trade Organization (WTO) have long been subordinate to the world's biggest financiers. They describe the WTO's (*the successor to the organization known as "GATT": the General Agreement on Tariffs [taxes on imports] and Trades, which was signed in 1947*) achievements as being, no more and no less than, a *"direct and dependent result"* of its member-state decisions and a dispute settlement system that commands mandatory jurisdiction (*Scholte and Robertson, 2007, p1309, 1310*).

While the WTO ostensibly exists to prevent worldwide cartels from *"strangling"* national interests (*and the nations of the world are persuaded to join and participate for this very reason*), the wily practices of those who are determined to plunder the entire world are slick and sophisticated. Its critics say that the WTO actually works hard to eliminate tariffs and strives to build legal and regulatory bases for unfettered, foreign, commercial engagements (*largely called "non-discriminating" trade practices*). As such, the WTO's internal dialogue arrangements appear to facilitate a *"quid pro quo"* approach to decision-making: *"You scratch my back and I'll scratch yours"*.

But, unfortunately, the poorest nations of the world cannot afford back-scratchers and are excluded from participating in important policy deliberations made, for example, by the Quad Group (*the US, EU, Japan and Canada*), the Group of Eight (*France, US, UK, Russia, Germany, Japan, Italy and Canada*), and the OECD. (*The reader will be pleased to know that the excluded developing countries, and the excluded less-developed countries (Australia is one of these), have opportunities to interact with and influence outcomes through the forums provided by the "Group of 77" and the "Group of 22".*)

It is hard to even be able to accuse such an organization of being callous, exploitative and self-interested under such inclusive, organized patterns of engagement which, superficially, appear to be inclusive and have generated tremendous prosperity for millions of human beings. But the current manifestation of the trade arrangements mean that the world's strongest and wealthiest decision-makers do not necessarily hear the voice of the world's most needy peoples. There is nothing new under the sun in this: those nations that have lacked both organizational aplomb and high monetary currency value remain poor and lesser developed because they have rarely been able to institute for effective local "*protectionism*" to manifest. Too many underdeveloped countries require the income derived from exporting resources to create a hand-to-mouth existence for their people; too many have corruptible authorities and individuals negotiating their trade arrangements.

Glob*ill*izers have promulgated the philosophy that "*trade under perfect competition maximizes aggregate welfare*". This mission statement, lauding the market ideal of competition, is now a rationalization being challenged strongly by responsible economists globally. It culminates, in John Ruggie's words, as providing for "*... little international capacity for redistributive transfers*" (*Scholte and Robertson, 2007,* p1314). If we are to have win/win forms of global trade emerge to end exploitation, we cannot abandon globalization's effective capacity to organize, achieve, develop and prosper. The organizational ability exercised is a tribute to human ingenuity, but ears must hear and alms must be given: we must, particularly on the nuclear electricity generating front, orchestrate for the missing "*redistributive transfers*". A simple, more inclusive, rearrangement of WTO groupings might be part of our way forward. The Group of Four and Group of Eight's members could commence to listen to close-by nations, those most in need of assistance and in concert with them, devise win/win terms of trade that engender prosperity and so, commence our millennium of replenishment. Creating electricity distribution networks and leasing fuelrods to enable industry and manufacturing potential, combined with desalination, will be part of the critical-to-achieve, generous contribution of infrastructure we now require to underpin a new era of cooperative win/win development scenarios for the entire world.

The global system of inequitable currency values also needs serious reappraisal. Because the Gold Standard fixed a huge variety of values for international currencies, foisting poverty on some, wealth on others, it is important, until fairer terms of trade emerge worldwide, that nation states, as much as possible, become economically self-sustaining and return to trading their surpluses and special goods for access to the products of other nations. I am sure, the recyclability of uranium will provide us with an opportunity to create the sense of *security needed for this fairness to emerge: there is plenty for all if it were to be compassionately and judiciously, shared.* Fuelrod leasing in exchange for industrial and handmade goods is quite feasible. Some of the most-valued artefacts worldwide come directly from periods of unsurpassed artisanship: the Baroque and Renaissance. India, in applying the wisdom of Ghandi, has certainly retained this capacity and many beautiful Medieval crafts are still widely practiced there. Returning to such skilled production and valuing that which is handmade is not beyond the potential of us all. It may be the route to follow to create opportunities for new, wealth-creating, global-market participation for the world's less-developed nations: it is certainly a well-travelled road, historically. Not a new thought: E.F. Schumacher's "*Small is Beautiful*" emphasized this particular, qualitative, course of action several decades ago.

The disproportional influence of lawyers, accountants and monetarists in a passively-invaded nation's organizational decision-making processes has indisputably helped to provide pre-eminence to the "*cost-rationalising (minimising) imperative*". Adherence to this particular principle has seen accountants appointed as directors and bureaucratic heads of departments and them being given the responsibility for overseeing all the organizations functions, including development. In these positions of responsibility they operate to singularly inflexible injunctions: "*reduce costs*" and "*operate within budget*"; "*adhere to the bottom-line*": catch-cries which became the economic rationalists' most-effective organization-manipulation tools! Because such materially-bound people (*recently called "eco-rats"*) control the purse strings, they are able to dictate an organisation's future activities and, so, enterprise can become strangled by book-balancing feats, not expansive vision.

This is a *topsy-turvy* and backwards decision-making, arrangement: the account-keeping process is subject to the vision, not the vision to accounts. Our futures are created by visionaries, those who are most gifted in knowledge and its adroit, creative, recombination. It is this gifted group of people, those rich in understanding the actual discipline of a department or organization, that are most likely to envisage infrastructural solutions able to generate future economic productivity, and it is those creative souls who are needed to lead the accountants, not vice-versa.

Throughout the entire, thirty-years plus of economic experiments that oversaw Australia's forfeiture of Government controlling national interests (*deregulating, floating the nation's currency, forfeiting national economic autonomy and independence by borrowing offshore, then selling infrastructure – privatization - and diminishing services to repay loans and balance cash-flow books*), horrid economic rationalism became the singular most important area of managerial accountability. Budgets were slashed, surplus trust funds stolen, research abolished, apprenticeships abandoned, career-mentoring stopped, workforces downsized and permanent appointees juggled into ill-matched positions, forcing the frightened remnant workforce to become compliant and ensure insecurity-based conformity to wrong, inefficient and nation-harming policies. Gone from the Public Service was the "*luxury*" of actually being knowledgeable and passionate about work for which one had a vocational affinity.

This episode has taken a heavy toll. Institute for second or third class results and that is the best you can hope for as an outcome: poor seconds and even poorer thirds. And so the nation suffers accordingly. Australia has had several decades of truncated government-based careers, awfully high local and national unemployment and distressingly high numbers of bankruptcies. All over the nation, the most profitable arms of Australian-owned enterprise and businesses, have been plundered and the nation's corporate, employment-providing, cost-minimising, tax-payer-funded, service-providing, bureaucratic structures have been dismantled, crippling the effectiveness of the residual public service.

As I write, several States battle bankruptcy, notably those that mortgaged state-owned assets to finance the Olympic Games and those that invested surplus superannuation monies in the international stock market a year or two before its collapse. The manufacturing sector is gone; our shipping is gone; most public infrastructure is foreign-owned: so quality-of-life opportunities are diminished. Interest rate rises are happening AGAIN, old-age pensions and services remain inadequate and many more cost-rationalized redundancies are threatened. The instability created by these changes fuel stock market volatility: something eco rats particularly love! If you can predict stock-market responses, you can make fortunes, but to predict stock-markets, you need to be in a position to manipulate them: only multi-nationals, whose annual budgets exceed that of whole nations, have that clout.

The end result of our abandonment of *"protectionism"* has been that several, successive, Australian Governments have been rendered largely ineffectual in both regulating, and providing security and prosperity, for, its own people. There are now beggars on the streets of almost every small and large city in Australia – there were none before this episode. All three tiers of Government have been affected and have become more and more dysfunctional. But having made their own *intelligentsia* redundant, they succumb to even greater dependence on offshore advisors and service providers, further entrenching the opportunities for passive invasion to continue its nation-plundering work. The entire has been an episode of disgraceful political mismanagement and betrayal of indigenous, black and white, Australians. Some have called the legacy of deregulated exploitation, *"the paradox of plenty"* to explain the frequently observed, inverse relationship between abundant natural resources and failing economic circumstances (see *Karl in Scholte and Robertson, 2007, V3, p858. He notes that this phenomenon has not only happened in Australia; many other nations, where similar tactics and policies were implemented, have also had their nation-state sovereignty, hugely undermined: including those from whom the policies emanated).* But I think it is more correct to reflect upon this episode as simply the most, recent, clever, way those who have little resources and much knowledge, and, much greed, managed and manipulated to obtain plenty for themselves. Such organizations are not likely to willingly forgo that which provides them with sustenance, so how to ensure that those from whom they derive their sustenance are not disadvantaged further is the challenge before us all.

While glob*ill*izers have a predictable, acquisitive, policy-making disposition, the better disposition for decisions to enhance a nation's development nearly always come from people who are prepared to assume responsibility for the well-being of their community, people who are trained in the discipline they are employed to contribute to, individuals who have rich local knowledge and who are familiar with the policies and practises of the past. When persons unaffiliated, and unconnected, to the historical community-of-people with whom they engage are placed in positions of influence, culminating in authority, their decisions are naturally going to reflect their own personal affiliations, knowledge and historic loyalties. Or they may simply reflect loyalty to an organization's bureaucratic procedures and salary provision *(selfish interest)*. So, in such situations, we have the adherence to convention, procedure, salaries and the cost-rationalising imperative triumphing over the characteristically-Australian, lateral-thinking, nation-building, determinations to organize and adapt situations to be able to be of genuine assistance to fellow Australians.

Decrying the practice of *"passive invasion"* is not a racist, nor anti-globalization, sentiment. The importance of diversity and welcoming newcomers is a well-established Australian trait, not only stemming from the ideal of *"mateship"*, but also from our cross cultural engagements with the world's most ancient, nomadic, people. Australia's Aboriginal people are strongly inculcated with both the importance of generously sharing to enable survival and the spiritual awareness of the impermanence of materialism. It takes great personal discipline and courage to live simply in a material world. They are a brave, ancient people. Over more than twenty generations of close contact, much of their cultural disposition has pervaded all elements of the white, indigenous Australians' cultural identity.

In fact, that the opposite of this sharing/caring, cultural disposition now widely manifests in bureaucratic and decision-making circles is how we Aussies can identify the invaders. Invaders are not particularly inclined to help fellow Australians: they simply do not see them as *"fellows"*. They will, instead, resort to rules and fail to adapt them to accommodate new circumstances or they may simply be, characteristically, only helping themselves, or more of their compatriots, into snug positions. They have no sense of being part of our continual, nation-creating exercise that is guided by providing a *"fair go for all Australians"*. Unhelpfulness is not the attitude with which this nation was built: mercenary, rule-bound, self-serving convenience and lack of helpfulness is decidedly un-Australian!

Deregulation's destruction

Deregulation affected the very certain-death injury, that which only a toxic, bitterly-poisoned, arrow could deliver, to the heart of *"protectionism"* and national autonomy. Putting the nation first was all that allowed Australia to generate independent wealth and actually finance her own defence when we were largely abandoned by the rest of the world in the first few years of World War II. Protectionist policies are a critical contributor to a nation's independent identity and essential for its autonomous prosperity: they are sane, sensible and will allow a nation to endure. Where there was full employment and nary a single beggar, and where all young families could aspire to own their own homes, Australia now has horrendous long-term unemployed and homelessness, couples defer creating families working under mortgages for years to try to create a family nest, while a cloud of continual, deliberately abolished security-of-employment hovers over all workers. Everywhere, short-term, casual, contracts have been created, endorsed and implemented by Government and by private enterprise. Even as late as 2010, today, a major political party's policy has announced more policies determined to minimize and abolish the conditions Australian workers enjoyed, those established decades ago and actively emulated by the rest of the developed world: conditions including weekend, public holiday and overtime penalty rates. More public service cutbacks are threatened. And, although the officially-approved interest rate is, as I write, nudging five percent, (*edging up from two percent over the months I have spent completing this writing project*) most of the private banks' rates are nearly double that amount. Increasing the cost of borrowed money is the primary inflationary force in an economy.

For forty years we have had nonsense ideas rationalised and then happen, like encouraging our young people to assume that they will have to become "*Jacks of all trades, masters of none*". This was, presumably needed, to provide the insecurely-employed with more chances of being able to respond to the awful vagaries of employment foisted on the nation. But, when coupled with the abolition of the apprenticeship training method for more than a decade *and* the sale, then dismantling, of Australia's manufacturing sector, these policies have only resulted in a massive deskilling of the nation.

Looking at employment opportunities now offered to Australians at home, it would appear that that the world's global economic engineers have successfully replaced the idea of "*quashing enervated masses through local warfare*" with the concept of quashing through manufactured unemployment and consequent subjugation to social welfare system that renders its "*dole*" recipients into a state of docile, resource-deprived, political impotence.

Recognising and lauding "*excellence*" also disappeared. It was replaced with thirty years of policies designed to encourage "*near-enough being good-enough*" outcomes at schools and in the Vocational Education and Technical (VET) sector. The system of education abandoned had strictly tested and graded all school-goers. Now mere *"passes"* are used to signify course completion in the VET sector: no grades kept beyond satisfactory or un-.

Education is the road to elevating one's life from menial servitude. It seems inexplicable that any political party sympathetic to the plight of working people could condone such changes; that they did is further evidence of Australia being effectively undermined in this quarter.

Educational policies that strove for, and nurtured, excellence supported Australians like Sir Marc Oliphant, the distinguished physicist who contributed to the Manhattan Project's development of slow neutron movement capture (*necessary for nuclear powered electricity development*). Oliphant's distinctions, commencing as a student at Unley and Adelaide High Schools, eventually won him a scholarship to attend the Cavendish Laboratory in Britain to continue his radiation and atomic energy studies. When he returned to Australia, he was a particularly active, anti-bomb, spokesperson, arguing that such a use was a *"misuse of atomic energy"*: a position with which still Australia concurs (*Elder, 1987, p396*). His career, like that of many other Australian nuclear physicists concentrated on developing non-military, useful, applications of the science. But without measures of excellence, it is possible that he, and many thousands of other talented Australians, would not have been identified and assisted to fulfil their potential.

We have looked at the fate of Australian universities, but have not mentioned that they were also downgraded by a political process that was distinguishing itself, not for progress, but for rekindling a largely-outdated *"clash of classes"* agenda with many politicians exercising a particular antipathy towards intellectuals. Barely literate Members who trust only superficial thought and read only simplified, spin-doctored, "*executive summaries*" to guide their policy initiatives. Who were these spin-doctors responsible for downgrading Australia's academic and educational institutes? Such downgrading "*flew in the face*" of the egalitarianism that was beginning to characterise the national Australian cultural identity emerging in the years that followed Federation and, in particular, characterised the energetic era of nation-building of post-World War II when the Australian National University was deliberately and successfully established to end "*the brain drain*" and assert independent academic excellence for Australia. The divisive and deliberate downgrading of academia eliminated the distinction between the various grades of instituted academic achievement, elevating technical colleges to university status. Coupled with deregulation, those policies are more of the "*fruit*", the evidence that shows how effectively Australia was both mismanaged and undermined.

From this, there is an important lesson to be learnt. When one is developing policy, one must test, both, the accuracy of the antecedent assumptions of its rationalizations *and* the relevance of the initiative by attempting to *predict the eventual outcomes of the policy*. Policy made without examination of these considerations is "*piecemeal*". This is why only the brightest minds should aspire to become political leaders. Without significant mental discipline and training, it is sheer opportunism to aspire to national leadership.

Understanding antecedent principles is part of the training one receives studying philosophy and classical logic. During this damaging, deregulating, episode, most of Australia's universities were forced to close their philosophy schools due to funding restrictions. The post-modernists' assumptions and conclusions culminating in the promulgation of ideas like "*situational ethics*" in the 1970s leading to "*greed is good*" in the 1980s, were, therefore, absorbed uncritically by many of the nation's decision-makers and the quality of the nation's objective and public discourse suffered accordingly. Rigorous ideological scrutiny was entirely replaced by "*spin doctored*" mass media contrivances, social engineering policy schemes and political positions designed to manoeuvre voting dispositions of either the majority, or the influential swinging voter, through adroit ideological positions promulgated by the mass media.

What I find most interesting and relevant to the theme of this book, is that, throughout these decades, most of the world's uranium was being precisely located. Aerial surveys, confirmed by on-the-ground geological survey data, revealed the comparative abundance of uranium across the ancient landscapes of Australia and on some other continents (*Africa, Russia, even The Tibetan Plateau*). This remarkable survey achievement and the discovery of comparatively abundant uranium reserves within Australia placed Australia in an economically unsurpassable position if she were to realise the ore's potential. But remarkably, instead, a rapid change of Government happened, which, combined with massive social mayhem, soon resulted in abandoning support for Australia's first nuclear-power electricity production plant. This incredibly, poorly-informed, fear-driven, decision was quickly followed by the gradual disabling of the Australian nuclear science community through career truncations and research funding squeezes. This happened because most Australians and their leaders neither understood fundamental nuclear physics nor, with their attention deflected to dealing with immediate, very probably engineered, significant, political crises, could they pay attention to learning/absorbing its importance.

Coincidence? No. This was not coincidence at all, but the contrived result of very successful "*passive invasion*" forces at work. I suggest that that entire forty-year episode of political and policy mayhem (*described in the Australian edition of this book*) can now only be viewed, with hindsight, as the deliberate creation of political, social and economic opportunities to exploit Australia (*amongst other nations*) of its invaluable, uranium oxide, (*and other*) resource(s).

We were caught off-guard: too few of us understood the value of the resource. Those that did were bound by the confidentiality and secrecy pertaining to all things nuclear since, even before, the Manhattan Project commenced. They were not allowed to speak to the press (*who, in ignorance, painted the nuclear science community as heinous dragons*) and Australia's nuclear science academy entered a hugely demoralised period, battling, frustrating, continual, funding cutbacks and abandonment of research opportunities. Several, inexplicable, sudden deaths robbed them of their most effective political advisors and spokesmen. The plundering of Australia's future became entrenched with contrived *pell-mell* conditions prevailing.

Glob*ill*ization was the solution the rest of the developed world applied in coming to grips with the exhaustion of their own local resources. Instead of pursuing replenishment policies at home, their attention followed the historic pattern humans have always employed, and they turned their energies to acquiring resources that could be readily plundered from further afield. Amongst the most sought-after acquisitions, uranium oxide being pre-eminent, were Australia's Government-owned essential services. They provided another, guaranteed form of income for the acquisitors. We, Australians, inadvertently allowed this plundering to happen because, here, a small population on the fringe of a large, dry, desert land actually had comparative plenty and we failed to be aware of how desperately hungry the rest of the resource-poor, developed world had become for precious resources.

The mayhem and "*economic rationalism*" that prevailed for forty years while globalization was allowed to gobble Australia, forcing bankruptcies, downsizing and redundancies on an unsuspecting, naive and genial people and resulting in significant service down-grading, particularly medical services for the most vulnerable of our people, the mentally ill, caused vast curtailment of collective prosperity and much personal grief. A callous, uncaring disposition had replaced nation-building sentiment and loyalty. Even today, potential protests against government divestment of public property are subjugated by fear of, and the actuality of, the protestors losing their livelihoods. The legislature-sanctioned whistle-blowing arrangements require from those who use the facility, confidentiality. This then precludes public scrutiny and debate over the many dreadful malpractices that have become commonplace in government bureaucratic places. Thus, the real political clout that public attention to malpractice can create becomes lost, minimising any individual or political accountability.

This period teaches us lessons of particular importance, both for Australia and to other nations seeking to prosper autonomously. The nation must come first: international trade needs to develop from internal productivity integrity; that able to be provided by a prosperous, self-sufficient national economy. Then a nation can trade its surplus and special, unique goods and services for the commodities it needs and wants.

Had the global nuclear-powered era, advocated here, commenced with the oil crisis of the 1970s, it is probable that the Australian manufacturing sector, powered by inexpensive electricity, could have remained an accessible and viable producer for the world's markets; the sector could have remained Australian-owned; and, with the likely, steady-growth, inexpensive electrical energy would enable, it could have guaranteed the continuance of the prosperity-creating full employment era for her people. The nuclear electricity producing utility prototype, approved for construction by the Australian Federal Cabinet in 1969, would very quickly have made all other electricity-generating utilities economically and technologically redundant. With inexpensive electrical power, our industrial output would have remained competitively attractive, despite our paying workers higher wages than many of our neighbours. Back then, *"Australian-made"* was regarded as equal to, if not, the best in the world.

But a reassertion of the political and economic influence of the fossil fuel oligopolies and the python-like grasp of multi-national take-overs combined to prevent this particular, much-needed, technological revolution from happening. Here, and all around the world, we exported, imported and burnt more and more fossil fuels. At the same time, nuclear power adoption was abandoned and banished in many other countries. The United States Government closed its commercial breeder reactors (*but kept their Defence installations operating*). A significant reason for this was probably that the breeder reactors would have made every other power generating utility in their country, economically uncompetitive. (*To be fair: it is probable that private, civil, operations able to produce plutonium (the intended product of breeder reactors) could have created the unacceptable risk of enabling weapons proliferation. Accordingly, the determination of the International Atomic Energy Agency to maintain fuelrod-manufacturing as a strictly-controlled Civic, not civilian, responsibility, enabling political accountability, military security and thorough bureaucratic scrutiny to prevail, is to be respected.*) The bloody, Gulf War happened, Indonesia invaded Timor and the deliberate, passive invasion of this particular uranium-supplying nation (*Australia*), *"upped its pace"*. Contrived, globalization (*gobbillizing*) and privatization policies usurped those that had ensured low interest loan rates and enabled and supported national economic, social and cultural autonomy and prosperity. One Prime Minister even announcing, with nary an effective appraisal or protest, that ours was *"a recession we had to have"*. Even the idea of Australian nationalism, that which united and propelled us to Federation, was deliberately replaced by the idea of "*multi-culturalism*". Multiculturalism was thought to be needed to help Australia defend herself against a hostile world. It was also thought necessary to end the residual, unjust, "*White Australia*" policy sentiment that prevailed in particular segments of Australian society as a consequence of the attempted Japanese invasion and a latent, real, fear of the "*dragon from the East aggressively rising*". But the concept of "*multi-culturalism*" shards and undermines the idea, and the actuality, of there being a unique, Australian, community of interest and cultural identity. In so doing, it discards Australian history and delimits our potential. Our identity is that founded on the unique, egalitarian, determinations given to Australia by the ideals of those who fought for nationhood using bare words and the gift of oration. Those solid words inspired and won the confidence of thousands of men and women who, nationless, laboured under, often, appalling conditions with no hope of bettering their prospects while they were deprived of political representation. The Australian cultural identity was developing unique, laudable, traits. It was a culture that valued loyalty to one's friends, "*mateship*". Mates were characters who, typically, were "*resourceful, tough, practical, sceptical, anti-authoritarian, and loyal*" (*see "Anzac Legend" in the Dennis et al, The Oxford Companion to Australian Military History, OUP, 2008, p37*).

Ideas that deliberately whittle away a nation's sense of self need to be carefully appraised to ensure they do not result in cementing hostile factions and entrenching potential disunity.

I say, "*move-over multi-culturalism*". While you have a place *everywhere* for facilitating tolerance and enriching greater of diversity, let it not be the pre-dominant place: better to have Australia known for being *a culture* that welcomes diversity and helps expands individuals' and communities' ability to experience a wide variety of satisfying cultural pursuits, than promulgate a multi-cultural *mish-mash* of competing loyalties. It seems important to return to taking pride in belonging to the place that lends to us its elements to create our unique beings: a place that provide us with the preparedness to accept "*new chums*"; a place that allows us to "*take the Mickey out*" of the pompous and tease the foreigners' accent without fear of creating trenchant animosity; and a place that provides all its people with the infrastructural support and ethical integrity that enables them to have "*a fair go*" at creating a wonderful life for themselves.

This same era of deregulating economic activity and imposing contrived economic hardship was philosophically underpinned by unscrupulous disregard for human beings potential (*see the collective writings of the Citizen's Electoral Council*) and an idea promulgated by the post-modernist school, that all "*truth is subjective*" and all ethics "*relative*": these fallacious doctrines used, and allowed, the idea of "*selfishness*" to be lauded as an inherent, inescapable, socially-acceptable trait. Our humaneness comes from our propensity to lovingly, even sacrificially, respect others, which in turn, creates the expectation of our receiving the same attention. Prostitution, even paedophilia were beginning to be rationalised as acceptable to proponents of this school of thought. So, the generous disposition that characterised Australians and their ideals of being "*mates helping and supporting one another*", were subtly eroded and replaced by conditions, here and in the rest of the Western world, that caused people to accept that they, selfishly and to survive, only had an obligation to "*look number one*". Two entire generations of young Australians have grown-up with these nation-harming rationalizations prevailing both politically and culturally. Meanwhile, Australia had become a quarry for globalization's gobblers. But, like *Artorius Rex* when the *Vandals* invaded his Welsh homelands, we are now finding ourselves "*fighting back*" to reclaim that which was precious, unique and inspiring to ourselves: the simple ability to create and nurture a distinctive community-of-interest from our own, unique assembly of people.

History of deregulation

Deregulation is reported to have begun in the developed world, in the United States of America, with the policies of Nixon, but became entrenched when Ronald Reagan's Government was persuaded by the world's biggest financiers to cut back family- and individual-support programs and reduce labour costs to, supposedly, *"improve profitability and encourage technological innovation and competitiveness"* (*Scholte and Robertson, 2007, p300*). Margaret Thatcher also followed Regan's ideology, and so untamed capitalism began to roar around the English-speaking, *"developed"* world, voraciously, unchecked, for an entire three decades. Keating, in Australia, followed suit. Part of this sordid story is described by a fast-growing group who now advocate an Australian Republic, The Citizens Electoral Council of Australia. I quote just a few sentences from one of their publications:

"...beginning 1983, first as Federal Treasurer and then as Prime Minister, Keating opened the country to takeover by foreign financial interests, by lifting Australia's exchange controls, floating the dollar, and dropping tariffs. These and additional globalization measures produced the worst foreign debt blow-out in the nation's history, from 38 billion to 206 billion, according to even the Keating Government's own, highly understated figures ...." (Allen Douglas, *Australian battle royal over "phony" republic*, Appendix A, p65 in *The Fight for An Australian Republic*, published by the Citizens Electoral Council of Australia, 2000.)

The *effective* legacy of *years of struggling* for a modicum of autonomy for the Australian economy to provide home-grown opportunities for individual and collective prosperity by our forebears was demolished by deregulation policies. In so doing, a mockery was made of the determination, the effort, the personal hardship, the ridicule and imprisonment endured by men and women who wanted Australia not to become embroiled in the wars, wrangles and economic woes that the old world sought to foist on them. The personal valour, the perspicacity, the brave assertions of leadership and the suffering wrought by persecution that brought our nation into being, was then rendered for naught. The ideals lovingly penned by bush balladists to unite and knit new groups of previously unaffiliated peoples in a deliberate determination to rally them to create a better place, a new nation they could proudly belong to and call "home", were discarded. Reverend Doctor Lang's decades of petitioning, forgotten, as was Curtin's courage when he was jailed for conscientiously objecting to World War I conscription. The sacrifice of the 60,000 men who lost their lives fighting World War I, shot in trenches in France and on the beaches and cliffs of Gallipoli: wasted. That, particularly tragic, event cemented the impetus that nurtured rapid, independent, development following the war because those who returned and those who remained were even more to determined to forge Australia into a strong, united, nation, independent of Old World, in her own right. Even so, and despite vehement protests, we have remained shackled to the wars that have followed. Contrast Keating's pandering to globe-gobbling ideologues with the strength of leadership shown by Curtain when he assumed responsibility for Australia as our Prime Minister in the Second World War when she was abandoned by Britain and his personal suffering when he decided to bring our troops home to defend our shores because of the high probability of them being discovered and destroyed by enemies during the sea voyage. Keating's nation-decimating policies also conveniently overlooked the dreadful privations and agony endured by the 15,384 members of Australia's 8th Division who were abandoned when Singapore was left undefended, then forced to labour with a huge proportion of them dying slowly by starvation (*either, while building the Burma railway or languishing in Changi Prison*). The tremendous ideological battles fought by those who wanted to avoid the impoverishment and woe caused by massive social inequalities and brutal repression based on the malfunctioning *"class"* distinctions that had prevailed in the Old World and paid workers a mere pittance, all of these efforts were totally forgotten by Keating (*if, indeed, he had ever been aware of them*) when he abandoned *"protectionism"* and laid Australia bare to the ravages of globi*ll*ization.

World War II's invasion of Australian shores happened at a time when, two years into the war, all our *trained* fighting men were stationed abroad and we had *"no tanks, no aeroplanes except for a few Wirraways, no pilots, and virtually no battle-ready troops"* (*find this riveting story by Robert Barwick, The economic mobilisation for World War II: Curtin's break with the British and alliance with America, in The Fight for an Australian Republic, op cit, pp52 – 61*). Under the leadership of John Curtin, and in a very short period of time, Australia became a modern industrialized nation, able to defend herself because Curtin called upon *"every human being in the country"* to work to this end. And they did: black and white. The monumental task of arming and equipping the nation for modern warfare was achieved under the technical direction of the brilliant of Essington Lewis, the man who had turned BHP into the *"best steel producer in the world"*. Australia's autonomy culminated when our Cabinet, in 1969, approved David Fairbairn's national development plans to institute nuclear power production, the result of many years of serious planning and research. At the same time we were prepared to assume nuclear defence capability (*hence forever removing ourselves from dependence on other powers thousands of miles away*). But this idea was then resoundingly hammered by one particularly vociferous politician who was *ag'in'* all things nuclear (*Tom Uren*). *"No nuclear power or weaponry"* has remained national policy since.

A particularly ironic outcome associated with the invention of nuclear weapons is that nuclear weapons operate most effectively by not being used: that is that they are a *"deterrence"* to war.

This development is perhaps the only *"good"* militaristic outcome that has manifest from the notion of possessing nuclear weapons capability. It is this principle of deterrence, based on the idea of a *"second-strike capability"* that is attributed to having ended both the Second World War and the Cold War, also preventing World War III between Russia and the United States and other more aggressive encounters between the world's super *(nuclear)* powers *(See Sheikh R. All, The Peace and Nuclear War Dictionary, 1989, p135, 178, 179).*

However, Australia's political decision to abandon anything *"nuclear"*, largely made on the grounds of our opposition to the carnage of *"nuclear war"*, is not a substantial enough reason for retrospectively understanding our having abandoned all things nuclear. No, unfortunately, abandoning nuclear electrical power development, and the capacity to defend ourselves by becoming a nuclear weapons power, straight away created an incredible dependence on military alliances abroad. It is those alliances that have dragged an unwilling and peace-loving Australia into all of her subsequent conflicts. It is they who have benefitted most from acquiring Australia's uranium inexpensively while the populace at large were left to absorb and harbour a deeply entrenched, incorrect and irrational fear of uranium's radiation, equating it with the horrors of the carnage of nuclear weaponry. This irrational, nationally-held, fear totally eclipsed the sensible idea of being able to use uranium's natural radiant energy for useful applications like electrical power production.

Deregulating opportunists upheld the simplistic idea of managing national economies by creating zero deficits and de-taxing the private sector. They entirely neglected to acknowledge the incredible, indispensable, function that **alone justifies the existence** of Governments, taxation and public services: publicly-owned infrastructure is needed to generate, significant, *future* economic activity and surplus wealth for the *entire* nation. Publicly-owned infrastructure and efficiently-provided essential services *enable* future wealth to be generated for both the maintenance of Government and establishment of private sector enterprises. A just, munificent, government is required to provide the conditions which will allow their entire people to thrive and prosper: that means investing in infrastructure to enable enterprises to commence (*and households to be run*) *with* minimum expense and so boost a nation's future economic activity capacity. Publicly-owned essential service provision and infrastructure creates employment, career streams, nurtures excellence in training and professional standards and eventually returns profits to the Government. In so doing, it helps reduce *all* taxes in the future. Publicly-owned essential services and infrastructure provide for an expansive economy. Australia's blinkered political leaders largely created the opposite: a contracting economy, devoid of essential service provision revenue and characterised by escalating unemployment. They then compensated for the ensuing lack of revenue by introducing the GST *on top of* payee taxes of 30 percent-plus, making Australians the highest taxed people in the world. Australians have endured this nonsense for forty years!

Retrospectively, the deregulation episode which gave *globillizers* unfettered opportunities to plunder the resources and enterprises of Australia amounts to treason. Treason to the lives given to building the nation, treason to the ideals they upheld and treason to future generations. It was an inept, callous and cruel, political episode, dominated by greedy opportunism and ignorance. It has revealed our political system to be grossly malfunctioning.

Fortunately for the world, this period is over. The deregulated systems **have not** delivered that which was used to justify their instigation: their promise of growth and more employment. All the deregulated economies now have huge public deficits, particularly Britain, Germany, France, Greece, Spain, the USA and Australia. Only in Canada and some Scandinavian countries, where the governments continuously, cleverly, invested in *"social policy, infrastructure and human capital programs"*, has economic stability and prosperity been consolidated for the majority (*Scholte and Robertson, 2007, p299*).

Reregulating for win/win globa*lli*zation scenarios

The challenge before us is how do we best redeem ourselves from the plight of having been so unknowing, gullibly, economically-subjugated? How do we replace globi*lli*zation with patterns of global trade that are beneficial to all of humanity? Is humanity capable of instituting for caring and sharing as efficiently as it has been able to institute to protect grasping exploitation? What subtle shifts in organizational and political policy are needed to enshrine and strengthen that which nurtures *the best of being human* in all of us and enable those qualities to triumph over that which is the worst in human beings.

Pragmatically-speaking, because the world's multinational globi*lli*zers have huge organizational impetus and financial clout, they are not likely to relinquish the economic and organizational systems they have established. If that were to happen, the economic repercussions would be severe. No, the remedial course of action required by humanity to prevent exploitation, and begin replenishment properly, necessarily involves *a deflection of their influence* rather than its direct curtailment.

Globalization's plundering ways are not irreversible. They have remedies. One such remedy requires us to use the economic activity they sponsor for *strengthening* national autonomy by actively committing to generating real wealth for the citizens from whom they have previously plundered. This can be attained by minimising offshore profit transfers and encouraging profitable investment in the countries they *"invade"*. Australia's Citizens Electoral Council, following *LaRouche* in the US and other economists, including the German Chancellor, also advocate a separation of investment banking from genuinely sustainable commercial banking. In so doing, the world's economic activity can be anchored to substance and removed from unsustainable, speculative gambling. At the same time, nation-states need to resume, at very least, resource apportionments of exported materials to enable local value-added economic development. And all nations need to *minimise,* to very small proportions of their budgets, private and offshore borrowings.

While we cannot easily, nor legally, annul existing trade arrangements and ownership profiles, we can most definitely ensure that the new nuclear–powered era is not subject to the same sort of invasive, grasping, manipulation that has endorsed *"beggar thy neighbour"* (an *AAP phrase*) trade engagements.

Both resource- and technology-providing nations need to establish mechanisms to ensure that a just share of raw resources, skills training and know-how, is made available so *all nations* can prosper. If, as the developed world uranium exploiters have argued, the uranium resource belongs to all people, then the technology and sciences that enable its use **also belong to all people**! A fairer form of globalization can be achieved by wealthy, developed, nations mandating, through United Nations bodies, perhaps the WTO; to provide access to uranium-based electricity-generating infrastructure, safely, inexpensively and as quickly as possible to all nations of the world. If the uranium ore body has the potential to provide the entire world with clean, safe, electricity production for thousands and thousands of years (*"Fermi's dream"*) from the outset of our planning, we are ***resource-enabled*** to make the ideal of greater infrastructural equity our global reality.

More-equitable global trade arrangements require us to institute arrangements such as:

a proportion of actual multi-national mined resources to be made available to the nations from whom the resource is taken;

creating opportunities for political skills development through mentoring, and allowing exchange representation, between both developing and developed country nations' managerial and governing bodies;

undertaking to train and teach sovereign nations' nominees the organizational operations and management skills that will allow them to co-manage Generation-IV and -V energy-production utilities safely;

providing nuclear-power fuel to all nations through fuelrod recycling **at a price their currencies and economies** can afford;

sponsoring appropriate undergraduate and graduate training in advanced technologies and management tools in *all* of the world's universities and colleges to enable the applications and opportunities that the uranium energy resource provides to manifest globally;

and, through a new win/win form of globalization, deliberately work to help create the conditions that allow the poorer nations of the world and the small island nations to apply the electricity and desalination potential to prosper their economy *and*, gradually, repair ecosystems to achieve genuine, ecological sustainability

In short: any form of global trade that takes raw resources (*minerals, gems, ore bodies, food, fish and forestry products et al*) from another, has a moral obligation to re-negotiate with responsible governing authorities and devise the most pertinent ways in which they can assist the providing nation to modernize and become prosperous. World trade organisers must now lend their management expertise to countries that have remained hungry and poor and devise mutually-beneficial ways for trade engagements to mature: this will be a shared capacity-building undertaking that may yield tremendous wealth for all involved. Providing a share of the company's produce and profits to the resource providers is the *most simple* form of reciprocal engagement. Many more productive and imaginative outcomes are possible when the directors of the multinational interests genuinely engage in conversations with developing nations' local communities and their governments to create innovative, geographically-appropriate, mutually-beneficial arrangements. Such respectful trade arrangements will allow the ancient ideal of *"noblesse oblige"* to permeate all global relations and create enduring and sustainable economies.

But even the realization of these ideals will not now save our planet unless the entire world commits to replenishing its bio-diverse ecosystems, and in particular, commits to replanting multi-structured, diverse, biomass-rich, forest. To achieve this feat will require all our governments' and globalizers' organisation aplomb *and* a great deal of money and manpower: amounts similar to that which the world currently commits to defence. There is no other pathway for us to take. Humanity has reached the end of the exploitive road we carved through the natural, living world, when, banished by our own self-indulgence, we fled the *denuded* Garden of Eden. We need to turn about, face our situation realistically and commence to repair the damage we have done. Our return to the Garden of Eden will only be achieved by our replanting and replenishing every single, desecrated place along the long road back. It is a journey that will take at least two hundred years to properly commence and a millennium to consolidate.

As the ideals of *"mutual obligation"* and *"noblesse oblige"* transcend the exploitative practices of globi*lli*zation's trade engagements, we have the hope that a great part of the cause of wars and terrorism will gradually abate. Those who are now politically and economically discouraged, dis-endowed and disenfranchised, will have the opportunity to reposition their nation's families of diverse people to create economic stability and prosperity if, in accepting that better relations and outcomes are possible, they can establish and maintain the level of organization, responsibility, integrity and inherent accountability required to lead all benevolent, just, governance arrangements. It might be that Australia may be able to help in unique way at this point in time. The Australian political system, based on division, must now quickly evolve to a higher form of democracy. We have seen, through the nuclear power story as experienced by Australians, that it has been possible to deceive an entire nation for one hundred years. Our ability to elevate our trade relations with the rest of the world would seem to depend upon our rising above remaining politically ensnared by popular, not wise, thought. To prevent this form of monumental duping from happening again requires Australia to institute an ability to provide governed people with leadership entrusted *and able* to make wise decisions on their behalf. In pursuit of this ideal, the idea of a *"meritocracy"* may have merit. This old idea was shared with me by a pre-eminent Northern Territory Musician, Ian Ellis, who has spent a lifetime pursuing excellence in performance and in his musicianship teaching. We do not like to accept *"mediocre performances"* in a concert hall, nor should we tolerate mediocrity in other public offerings, particularly in our political arenas and governance arrangements.

The meritocratic ideal

A *'meritocracy'* is instituted when competence, truth, excellence and high quality outcomes are the ideals upon which decisions are made and performance is ranked. It is formally defined in the Australian Oxford Dictionary as *"government by persons selected competitively according to merit"*. (Note: merit is the selection criteria, not whether the candidate is liberal, labor or

green, black or white: such political conglomerates have created
an Australia that is artificially divided by political strife.)

A *meritocracy* does not allow the lowest common denominator
of popular thought to determine state or national policy, such as
would happen in a *plebiscite,* a referendum, or when caucus
conglomerates play *"numbers games"* to cement, factionalized,
party policy. It requires independent, meritorious, candidates,
those whom their entire electorate can trust to represent their
unique, collective, community of interest, to stand as candidates
and replace the now redundant, liberal/labour,
republican/monarchist, divides.

It is rather unfortunate, but history does not hold the opinion of
the *'pleb'* as something inspiring: the vote of the *'plebian
assembly',* has, since ancient times, been thought to be
'ordinary', 'insignificant' and *'common'.* A plebiscite about
nuclear power in Australia, at this point in time, is likely to see
this rational, good, idea soundly defeated and perhaps defeated
for many years to come. Does Australia, or does, indeed, the
world want her nations to be bound to *'ordinary', 'insignificant'*
and *'common'* decisions? Or would we like to aspire to being the
best we possibly can be and be led by our most knowledgeable
and responsible individuals?

Leaders in a modern meritocracy seek advice from those who are
qualified to give opinions and, because of their commitment, love
for their people *and their intelligence,* the leaders (*and their
advisors*) are entrusted to make the right, not necessarily the
most popular, decisions. In making *"right"* decisions they are
obliged to be respectful of all, securing the well-being of all their
electorate's people. To achieve this, they must be required to be
mindful of the advances we have made in establishing universal
human rights and not create harm for others. Following such
precepts, they would be obliged to nurture opportunities for

universal education and full employment so prosperity, through the provision of fair, generous, wages to all workers, can happen.

It also seems appropriate to suggest that a meritocratic parliament be overseen by a group of pre-eminent and knowledgeable, especially distinguished, long-serving, individuals. Individuals: not a single President who may tire, or may be easily diverted by influences that are not in the nation's collective long-term interests. The overseeing individuals should be capable of appraising the decisions made by meritocratically-elected representatives, assessing their petitions and able to advise and authorize expenditure for development accordingly. In Australia, such a group of people could be called "*the nation's elders*" out of respect for the first Australians, our Aboriginal people, and our Senate could accommodate this assertion of responsibility. Little, if any, constitutional change would be required to implement such a system.

If Parliament were to be instituted in this way, as a **meritocracy,** it would more closely correspond with **the natural way humans relate**. We need to be led by our brightest and most able. Families have their head of the household, the father or mother, grandfather or grandmother, who listens to each member's requirements, appraises their relevance, then makes decisions in their family's best interest; villages have their chiefs who, if they fail to look after their community members, find themselves ousted from the chiefdom; countries have either their chief ministers or their kings and queens and House of Lords. All of these elevated-personage roles *can* serve to inspire us all to achieve greater and better things because, ideally, their intelligent actions are worthy of emulation. Looking around the world, we find that leadership is sometimes conferred through inheritance (*because the inheritors are provided with the best possible educations and trained from a young age to assume responsibility for the well-being of others*) and, in other societal arrangements, it is conferred through recognizing the manifestation of excellence and networking competence. If those who assume, or are assigned to, leadership positions become corrupted and commence to represent self or particular vested interests, then despotism will reign and the well-being of all is likely to be truncated.

One measure of a truly competent leader is their ability to share their insights and reasons *to elevate public opinion* to support the positions they hold to be most beneficial to our nation and not let their people remain in a quagmire of ignorance. As such, meritorious leaders must not simply repeat that which might be manufactured for them by spin doctors, delivered through television speech prompts. Genuine leadership embraces and must deliver pinnacle thought and performance, simply and eloquently.

In Australia, we do not have inherited governors. Rather, we are entitled to choose our leaders through voting, an exercise in fundamental democracy. Wisely exercising this choice requires us to assert value judgments about the candidate's ability to represent our individual and collective interests. But in Australia the pool of suitable candidates has been circumscribed by all of our political parties' preoccupation with looking after their party membership's lopsidedly lobbied positions. If only a tiny percent of the nation belong to either the party or to unions, is it fair that that tiny percentile should determine national Government policy? If that party membership represents very wealthy offshore cartels or workers unions, requiring access to more of our coal, gas, gold, telecommunication resources and other infrastructure - groups who are willing to pay a huge amount to help their targeted party win the next election - is it right that they be granted access to those resources and the rest of Australia be denied the opportunity to be assisted to value-add and generate local and public prosperity from harvesting and developing the same resources? I think not.

To circumvent this problem, we require courageous, compassionate, meritorious, citizens to stand as candidates, men and women capable of genuinely representing their entire electorate and its development needs. Electoral offices could then be reinvigorated to become venues for genuine exchanges of ideas, full and occupied with listening, appraising, negotiating and planning the best outcomes for all; not forums for power grabbing by a few.

To conclude

Ideals are just ideas until they are put into practice. Australians are fortunate that, by virtue of the abundant uranium ore body our ancient landscape has endowed to the continent and because her people are innovative, we can immediately commence to build a new generation of technology that allows a re-assertion of national sovereignty to guide our future energy production infrastructure, transport, communications and water distribution efforts. We are blessed in that we have the resources and the technological skills to enable us to transcend the manipulation that resulted from our having thrown our economy to the vagaries of deregulation and offshore manipulation. The episode of privatizing our essential services and allowing unfettered foreign access to plunder the nation's mineral, agricultural, forestry and energy wealth, without seriously replenishing and apportioning to meet our own future needs, is now over. As such, Australians who participate in the global economy are in a very good position to immediately help construct and instigate globalization organization and trade arrangements that insist upon win/win economic arrangements with the rest of the world. When we achieve such organizational arrangements here, we will be in a strong position to help our immediate neighbours, where some of the world's poorest people live, to achieve the same. To do so would be a re-assertion of the values of men like the Reverend Doctor John Lang embodied, enacted and taught to help knit together the diverse representation of people we now know as Australians, into a free and prosperous, independent nation. We now also know that when Lang stood and spoke to Australians and asked them to unshackle themselves from the past and its injustices, he was not only speaking to us, he was speaking to the entire world, rallying one and all to the justice and sense of creating national unity and prosperity through us all assuming ethical, sovereign-nation responsibilities.

How much more civilized humanity will become when collectively we decide to leave a legacy of generosity, benevolence and efficient systems and infrastructure from our global trade arrangements - those that enable prosperous development - rather than the trauma wrought by the morally-bereft, resource-grasping, war-mongering that has characterized a huge proportion of more voracious and mobile nations' engagements with others throughout human history.

For peace-loving Australians, helping our neighbours to attain similar prosperity will be one way of redeeming ourselves from having committed to wars against people who were not our enemies.

It is the only way we have to redeem our tattered international intellectual reputation stemming from our having been so very thoroughly "*duped*" on the issue of the value of uranium. For, it is arguable, there has never been a plundering incident in history that can equate to that which has happened to the Australian people in the instance of its uranium ore body: a value realized in the early 1900s but which, thereafter, was very carefully hidden for one hundred years by ore-plundering, *passive invaders*.

We have the hope that if the lessons learnt from the Australian experience can be made available to the rest of the world, other nations, in turn, may commence to minimize the manifestations of similar exploitation and manipulation. This objective is mindful of the ominous deterioration in the life-supporting potential of the biosphere. We face an unrelenting need for us all to, immediately and effectively, take responsibility for the biosphere's well-being and actually apply the organizational skills needed to enable us to effectively halt, and reverse, climate change.

Here, in Australia, our sense of having been unjustly dealt with is fuelled further by the knowledge that, for at least as long as sixty years, possibly one hundred years, the most advanced pockets of global intelligence have known that nuclear power would, and should, replace fossil fuels. But sharing this knowledge was delayed, not only to protect existing oil and infrastructure investors, but also to provide the knowledgeable few with the opportunity to reposition their nations to gain a plenitude of uranium oxide. We also now know that if we had, globally, collectively and co-operatively, acted forty years ago to institute nuclear-powered, alternative, energy-generating capacity (*as Australia and others then planned to do), a vast amount of the damage* caused by our continuing to burn fossil-fuels could have been averted. And had we implemented this infrastructure, nuclear power would now cost *a fraction* of current electricity generating costs.

Those few reflections about the dynamics and politics of energy production over the last century are all we need to propel us into immediate action to demand, encourage, enable and institute, far more responsible international engagements. These lessons are shared with the view of helping to enable smaller populations, like ours, all around the world, to begin to reposition their own national energy policies more responsibly and appropriately.

The advanced scientific knowledge and amazing levels of sophistication and civilized behaviour now attained by the most developed nations of the world could be lost to all of us forever if the life-supporting potential of the biosphere collapses further. We shall come perilously close to that eminently, unsavoury, scenario if the developed world's advances in knowledge and organizational ability are not shared to enable the entire human family elevate its circumstances to similarly advanced positions. Humanity must now implement a huge, co-operative, globe-encompassing, organisational effort to have any hope of restoring and replenishing global ecosystems, especially the forests of the world. Indeed, we must immediately *"beat our swords into ploughshares and our spears into pruning hooks"*. That quoted, particular, sweet, vision of our future, shared thousands of years ago by the prophet Micah, is even richer: it acknowledges our need to be led by ideas and God-given, higher ideals, and it reminds us to respect, fundamentally, the diversity and differences that are an integral part of the human family today, and always, for he continued:

"Nation shall not lift up sword against nation, neither shall they learn war anymore. But everyone shall sit under his vine and under his fig tree, and none shall make them afraid. For the mouth of the Lord of hosts has spoken. For all people will walk in the name of his god, but we will walk in the name of the Lord our God for ever and ever and ever." (Micah, 4: 3-5 NKJV)

Making this tolerant, gracious, divinely-inspired injunction, our reality, is the superlative challenge every community, race, nation and organizational affiliation now faces. With the management techniques, the communication ability, the wealth of knowledge and the technological capability we humans now possess, and with such a splendid vision to guide us, hope abounds that Micah's blessed ideal will soon become our collective, global, reality. Ends June 2010.

References

A Clarion Call: Last Stop Before Chaos, The Sydney Morning Herald, November 1st, 2006, article appraising *The Stern Report,* http://www.smh.com.au/news/environment/a-clarion-call-last-stop-before-chaos/2006/10/31/1162278141640.html

ANA Conference, 2007, *A Nuclear Future, 7th Nuclear Science and Engineering Conference,* Sydney Mechanics School of Arts, 19th October, 2007. Proceedings available through Australian Nuclear Association, PO Box 445, Sutherland, NSW, 1499. ISBN 978 0 949188 16 8.

Australia's Energy Options, Future Directions International Study, October, 2005.

Australia's Uranium: Greenhouse Friendly Fuel for an Energy Hungry World, November, 2006. The Parliament of the Commonwealth of Australia, House of Representatives Standing Committee on Industry and Resources, Chair: The Hon Geoff Prosser MP.

Boyden Stephen, 1990. *Our Biosphere Under Threat,* Oxford University Press.

Boyden Stephen, 1992. *Bio-history: The Interplay Between Human Society and the Biosphere, Past and Present, UNESCO* and Parthenon Press.

"Burning" Weapons Plutonium in Candu Reactors, verbatim excerpts from Management and Disposition of Excess Weapons Plutonium. Committee on International Security and Arms Control, US National Academy of sciences, 1994, access date: 11th September 2008. http://www.ccnr.org/nas_mox.html

Can we Burn Plutonium?, New Scientist, reprinted from 3rd Jan 1957, accessed 11th September, 2008 from 50 Years of New Scientist: The Best Articles.
http://www.newscientist.com/channel/opinion/classic-

articles/dn10526-can-we-burn-plutonium.html?feedId=classic-articles_rss20m

Citizen's Electoral Council of Australia, *The fight for an Australian Republic from the first fleet to the Year 2000, May 2001.* Citizens Media Group Pty Ltd 136 – 144 Bell Street, Coburg, Victoria, 3058.

Cohen BL, 1990. *The Nuclear Fuel Option: An Alternative for the 1990s,* Sage Publications Inc, 233 Spring Street, New York, London.

Daly F, 1982. *The Politician Who Laughed,* Hutchinson Group (Australia) Pty Ltd, Cremorne Street, Richmond, Victoria.

Davis D, (Devra) 2007. *The Secret History of the War on Cancer,* Basic Books, a member of Perseus Books, Park Avenue, New York.

Dennis, P, Grey, J, Morris, E, Prior, J 2nd ed. 2008, *The Oxford Companion to Australia's Military History.* Oxford University press, South Melbourne, Victoria.

Elder, B, 1987. *The A to Z of Who is Who in Australia,* Child and Associates Publishing Pty Ltd, Brookvale, New South Wales.

Halisham, Lord, Quentin Hogg, 1976, *The Elective Dictatorship,* Richard Dimbleby Lecture, BBC, 1976, see extract: http://law.uts.edu.au/~chrisel/hailsham.html

Hardy CJ, 1996 & 2nd edition 2008. *Enriching Experiences: Uranium Enrichment in Australia 1963 – 2008,* Glen Haven Publishing, PO Box 85, Peakhurst, NSW, 2210.

Hardy CJ, 2006. *A Cradle to Grave Concept for Australia's Uranium,* Lecture to the Annual Meeting of the Four Societies,

Engineers Australia, 22nd February 2006 Copies: Australian Nuclear Association, PO Box 445, Sutherland, NSW Australia, 1499.

Jackal R, 1980. *Structural Invitations to Deceit: Some Reflections of Bureaucracy and Morality* in <u>Berkshire Review</u> (15): 49 – 61.

McGaffin W & Knoll E, 1968. *Anything But the Truth: The Credibility Gap – How the News is Managed in Washington*, New York, London, McGraw Hill.

National Archives, <u>www.naa.gov.au</u>, see David Fairbairn's Correspondence, and Paper of the Series AN976/642 , <u>http://naa12.naa.gov.au/scripts/provenance_lising.asp?F=1&5=6&c=11</u>

Pope, S & Wheal, E 1995. *Introduction* by Stephen Pope in *The Dictionary of The First World War*, St Martin's Press, Scholarly Division, 175 Fifth Avenue, New York.

Register of Australian Mining 2009/10, published by Resource Information Unit, 79 Hay Street, Subiaco, Western Australia.

Sheikh R. All, 1989. *The Peace and Nuclear War Dictionary*, ABC-CLIO Inc, PO Box 1911, Santa Barbara, California, USA.

Scholte, Jan Aart and Roland Robertson, editors, 2007. *Encyclopedia of Globalization*, four volumes. MTM Publishing, Routledge,445 West 23rd Street, New York, NY 10011, <u>www.mtmpublishing.com</u>

Taylor L. <u>Weekend Australian</u>, Sept 6, 2009. *Funds short for solar ambition.*

Williams David R, 1998. *What is Safe? The Risks of Living in a Nuclear Age*, Department of Chemistry, Uni of Wales, Cardiff, UK.

Appendix one
Useful terms

These terms have been compiled with reference to four sources: Longman's *Dictionary of Environmental Science*; Elsevier's *Dictionary of Energy*; the glossary provided in *Australia's Uranium – Greenhouse friendly fuel for an energy hungry world; and* from Larkin and Peters, *Dictionary of Concepts in Human Geography.*

Acculturation refers to the assimilation or replacement of cultural traits by absorbing those from another culture.

Afforestation differs from reforestation which is planting replanting forests on sites they formerly occupied. It is actually establishing new forest on land that has not been treed in our recent geological past.

Aposteosis the natural expulsion of a malfunctioning, damaged or dead cell from the body.

Biosphere refers to the entire living world including the Earth's atmosphere and hydrosphere (oceans, rivers and rainfall).

CANDU the name of a pressurized, heavy-water reactor developed in a partnership between the Canadian Atomic Energy Commission and the Hydro electric Power Commission of Ontario in the 1960s. It was primarily designed to use un-enriched uranium, but can operate with mixed oxide fuels.

Carcinogen a substance able to cause cancer.

Conservation of the natural environment serves the purpose to preserve species, genetic and community diversity and biomass.

Critical load the maximum sustainable use of a resource, or, the maximum amount of pollution, after which, if this threshold of a "*critical load*" is reached, the pollutant's effect is not able to be nullified by natural processes.

Ecological footprint the impact a city or nation of people has on the surrounding natural world, referring especially to resource depletion. Minimal footprint, not exceeding any critical loads, will enable human societies to become ecologically sustainable.

Ecosystem simplification refers to the inevitable chain-reaction of destruction that is associated with the disappearance of a species, a community or ecosystem. Simplification is then inevitable because relationships between living entities are co-dependent and symbiotic in nature (*see the "Feature essay" at www.specialistwritingservices.com.au*). Conservation seeks to arrest and reverse this process.

Ecological sustainable development Ecologically sustainable development requires humanity to minimize damage to, preserve and protect, restore and replenish, bio-diverse biomass to enhance nature's ability to self-perpetuate life through supporting increasingly diverse interrelationships between ever-diversifying, fluctuating communities of co-dependent species. These interdependent communities, together with their legacy of modifying the physical parameters of their local environments, help to create the ecosystem functions that collectively create the conditions we can describe as "*a biosphere amenable to the evolution of life*".

Energy conservation   to conserve energy means that the absolute amount of energy used for a particular purpose is reduced to being equal to the exact amount required, and/or, that waste energy is minimized.

Energy efficiency   measures the amount of energy required for the job at hand compared to that which is actually expended in performing a particular task. High efficiencies mean that the difference is small: its waste energy (*mostly heat*) is little. Low efficiency means that there is a lot of input energy for the achievement of little work with much waste; hence room for the implementation of energy conservation

Fertile material   is that which is receptive to capturing neutrons from fissile material. It may transmute to another substance, as does uranium-238 in becoming plutonium; it may produce stable or unstable isotopes following neutron absorption.

Fissile material   is material which naturally, or when significantly heated, alters its internal nucleus composition, splitting and, in the process, releases energy and neutrons. Uranium-235 is naturally "*fissile*".

Fusion describes the process common to the formation of all elements. At very high temperatures, light-weight elements combine, "fuse", to produce heavier elements. Hydrogen's internal atomic composition would appear to be the basic building block for this process. For example, two hydrogen atoms fused create helium.

Geothermal energy   the natural heat energy of subterranean magma fluids, rocks and fluids produced by the naturally radioactive elements of thorium, potassium and uranium in Earth's core.

High-grade   generally relates to quality and accessibility issues, that which is *'high-grade'* is the best and the least expensive to process. **High-grade energy** is that which provides abundant, reliable, readily convertible energy for the job at hand *and* is relatively inexpensive to access or harvest.

Hormesis   describes the healthy *"tonic effect"* that low level radiation exposure, over a long period of time, can provide.

Integrity   *"Integrity"* is an interesting concept. It means to be honest in one's undertakings. I also use the term in relation to how closely people's positions or policies portray a **truth-based** modern, accurate, paradigm.   Truthful paradigms are supported, or discredited, by established facts and knowledge-derived, respectful of human rights, wisdom. For example, supporting fossil-fuel burning adheres to an outdated paradigm: this is an unsustainable, atmosphere-harming practice. Advocating *"clean carbon"* technologies relies on incorrect assumptions and it is not energy-use efficient.   Clearing forest, without respecting sustainable-harvest principles, adheres to a plundering, damaging, greedy paradigm and ignores respect for interconnectivity of life, mutual dependence. It perpetuates the destruction of the forest-derived, crucial to life, climate-stabilizing-function bio-diverse biomass provides to the entire planet.   A politician who advocates these practices, instead of facilitating transitional aid to that which is sustainable, has very little integrity because the paradigms they serve are not only dysfunctional and outdated: they are harmful, hence, immoral.

International Atomic Energy Agency, IAEA, established under the auspices of the United Nations in 1957 to promote the peaceful use of nuclear power.

Isotopes   refer to variation of atoms of the same element which possess slightly different mass (*heavier or lighter*) properties to the most prevalent form of the atom.   A

radioactive isotope is comparatively unstable form of an element and emits energy in pursuit of internal stability.

Megawatt (MW) is a measure of electricity equivalent to one million watts: the amount needed to light 10,000 globes issuing 100-watts. A single **Watt** describes the transfer or conversion of energy, one joule, over a unit of time, one second.

Modernization prefer Learner's definition, "...*current term for an old process, the process of social change whereby less developed societies acquire characteristics common to more developed societies*".

Natural gas exists either as a gas or in solution with crude oil. Most often it is separated from oil through processes which include the heavier hydrocarbons being absorbed or forming a condensate. Liquid natural gas, LPG, LNG, are generic names applied to many grades of crude oil. A modern euphemism for "*oil refinery*" is "*LNG plant*".

Naturally-occurring gases are those simple molecules that form gases at the temperatures and pressures that prevail at the Earth's surface. They include the gases formed from combinations of carbon and hydrogen (*hence, hydrocarbons like methane, propane, butane, ethane*) and heavier non-hydrocarbon compounds (*nitrogen, hydrogen sulphide, carbon dioxide*).

Non-renewable acknowledges that a resource, degraded, removed or used, is not able to be replenished in human civilization's time frames (*but not necessarily in terms of geological time*).

Nucleus (nuclear) relates to the strongly bonded, dense, centre of an atom.

Nuclear energy refers to the energy derived from the fission (*splitting*) or fusion (*fusing together*) of two or more atomic nuclei.

Petrol/gasoline refined petroleum.

Petroleum   a collective term for crude oils, natural gas, natural gas liquids and other hydrocarbons.

Photosynthesis is the process by which green plants are able to use the sun's energy to convert carbon dioxide and water into more complex biological chemicals which provide chemical energy for more life (*plant growth*), which, in turn, is able to be used to provide energy, as food, to other forms of life. A byproduct of this conversion is oxygen, which diffuses into the atmosphere as a gas.

Photovoltaic   is a term applied to the capturing of the sun's electromagnetic energy and its conversion into electricity by use of photovoltaic devices or cells manufactured from compounds including silicon, arsenic, cadmium, chromium and tellurium.

Politics (*conventional*)   is a form of political activity defined as "*the pursuit of power to exercise control*" and in George Blahusiak's words: "*A definition that does not say anything about honesty, integrity or good government and it certainly doesn't say anything about benefitting the governed*" (*pers. comm. November, 2009*). Machiavelli's tired and cynical appraisal of successful politics included that "*the means is justified by the end goal*". But no end can be assured: the future will always be influenced by the unanticipated and unexpected. So, to have genuine political integrity one must necessarily pay very strict attention to the means used to reach a desired goal by the application and institution of the best human rights can avail to us, not the worst. Hence, **Politics (*21st Century*)** seeks to abandon the, tired, unprincipled, Machiavellian definition and redefine politics *"as the pursuit of distinguished organization manifesting as excellent outcomes for all governed peoples, and their natural environment, through wise decisions made by caring, accountable, leadership committed to the ideal of the "Golden rule".* (*Dr Bernardine Atkinson, November, 2009*).

Radium is a radioactive, metallic element that is usually found in uranium ores and substances like pitchblende. Radium-226 has a half life of 1602 years and is useful in radiation therapy and research. It decays to form the noble gas, radon.

A refinery employs the processes of distillation, cracking (*a process used to decompose, break, complex carbon compounds into smaller molecules*), purifying and other treatments to transform crude oil into commercial products including helium, hydrogen, propane, butane, methane and aviation gas. Some of the heavier *"fractions"* are used to fuel these refining processes and, thereafter, cool the gases to the point of *"liquefaction"* to enable their transport under low temperature (*minus 260°F*) and high pressure conditions: energy-intensive processes.

Royalty is a payment made for access to, and use of, property belonging to another.

Sequestrate means to *"store"*.

Sialic crust materials these are Earth's oldest, non-sedimentary rocks, mostly granite, found on the continental crust. They are aged as 1,500 million years old.

Transmutation describes a nuclear reaction that changes one element into either a new element or a new isotope. When super-heated, radioactive waste can be rapidly transmuted to nonradioactive elements.

Transpiration   the evaporation of water (*water vapor*) from the leaves and stems of plants.

References

Cleveland, CJ; Morris, C 2006, *Dictionary of Energy*, Elsevier Ltd, The Boulevard, Langford Lane, Kidlington, Oxford, OX51GB, UK.

Larkin, Robert P and Peters, Gary, L *Dictionary of Concepts in Human Geography*, 1983, Greenwood Press, 88 Post Road West, Westport, Connecticut, 06881.

Lawrence, E; Jackson, ARW; Jackson, JM 1998 *Longman Dictionary of Environmental Science,* Addison Wesley Longman Ltd, Edinburgh Gate, Harlow, Essex CM20 2JE, England.

Australia's Uranium: Greenhouse Friendly Fuel for an Energy Hungry World, November, 2006.   The Parliament of the Commonwealth of Australia, House of Representatives Standing Committee on Industry and Resources, Chair: The Hon Geoff Prosser MP.

Oxford Dictionary of Physics, 5th edition, 2005.   Oxford University Press, Great Clarendon Street, Oxford, OX2 6DP.